The
Platelet Ami
Storage Gra

The Platelet Amine Storage Granule

Edited by

Kenneth M. Meyers • Charles D. Barnes
Department of Veterinary and Comparative Anatomy,
 Pharmacology and Physiology
College of Veterinary Medicine
Pullman, Washington

CRC Press
Boca Raton   Ann Arbor   London   Tokyo

Library of Congress Cataloging-in-Publication Data

Catalog record is available from the Library of Congress.

Developed by Telford Press.

International Standard Book Number 0-8493-8838-4

Printed in the United States
1 2 3 4 5 6 7 8 9 0

Printed on acid-free paper

CONTENTS

CHAPTER 3

CHAPTER 5

POLYPHOSPHOINOSITIDE METABOLISM IN RESTING
AND STIMULATED PLATELETS

ABOUT THE EDITORS

Kenneth M. Meyers, Ph.D., is Professor of Physiology and Pharmacology in the Department of Veterinary and Comparative Anatomy, Pharmacology, and Physiology in the College of Veterinary Medicine of Washington State University at Pullman.

Dr. Meyers attended Montana State University and received his B.S. degree from Oregon State University in 1964 with a major in zoology. He received a Ph.D. degree from Washington State University in 1969. He advanced to Associate Professor in 1974, and in 1979 he became Professor of Physiology and Pharmacology at Washington State University. In 1975 and 1981 he was a visiting scientist at the Center for Thrombosis Research at Temple University.

Dr. Meyers is a member of Phi Zeta, Sigma XI, American Society of Pharmacology and Experimental Therapeutics, Hemostasis, American Association of Blood Banks, and the American Association for the Advancement of Science. He has been the recipient of grants from the National Institutes of Health, American Heart Association, and the Washington Heart Association.

Dr. Meyers is an author of more than 70 manuscripts and 6 book chapters. He is currently a recipient of a Transfusion Academic Medicine Award. His research centers on control of platelet function in health and disease.

Charles D. Barnes, Ph.D., is Professor and Chairman of the Department of Veterinary and Comparative Anatomy, Pharmacology and Physiology in the College of Veterinary Medicine at Washington State University.

Dr. Barnes received his B.S. degree from Montana State University in 1958 with double majors in biology and physics. In 1961 he received an M.S. degree in physiology and biophysics from the University of Washington, and in 1962 he earned his Ph.D. in physiology from the University of Iowa.

After two years as a postdoctoral fellow in the Department of Pharmacology at the University of California at San Francisco, he became an Assistant Professor of Anatomy and Physiology at Indiana University in 1964. He advanced to Associate Professor in 1968, and in 1971 became Professor of Life Sciences at Indiana State University. In 1975 he was named Chairman of the Department of Physiology at Texas Tech University College of Medicine, where he remained until taking his present position in 1983.

Dr. Barnes is a member of the American Association for the Advancement of Science, American Association of Anatomists, American Institute of Biological Sciences, American Physiological Society, American Association of Veterinary Anatomists, American Society of Pharmacology and Experimental Therapeutics, American Society of Association of Veterinary Anatomy Chairpersons, Association of Chairmen of Departments of Physiology, International Brain Research Organization, Radiation Research Society, Society for Experimental Biology and Medicine, Society for Neuroscience, Society of General Physiologists, and the Western Pharmacological Society. He has been the recipient of many research grants from the National Institutes of Health and the National Science Foundation.

Dr. Barnes is the author of more than 150 papers and has been the author or editor of 15 books. His current research interests relate to the modulation of nervous system output by centers in the brainstem.

CONTRIBUTORS
(IN ORDER OF CHAPTER PRESENTATION)

JAMES G.WHITE, M.D.
Regents' Professor, Associate Dean for Research
Laboratory Medicine and Pathology, Pediatrics
University of Minnesota Medical School
Minneapolis, Minnesota 55455

MICHELE MENARD, D.V.M., M.S.C., Ph.D., DIPLOMATE ACVP
Assistant Professor, Veterinary Clinical Medicine and Surgery
College of Veterinary Medicine
Washington State University
Pullman, WA 99164-6610

KENNETH M. MEYERS, Ph.D.
Professor, Department of Veterinary and
     Comparative Anatomy, Pharmacology
     and Physiology (VCAPP)
College of Veterinary Medicine
Washington State University
Pullman, WA 99164-6520

HOLM HOLMSEN, Ph.D.
Professor, Department of Biochemistry
Preclinical Institutes
University of Bergen
Årstadveien 19, Bergen N-5009
Norway

KAMIL UGURBIL, Ph.D.
　　　Professor, Biochemistry/Radiology Medicine
　　　Center for Magnetic Resonance Research
　　　University of Minnesota
　　　Minneapolis, MN 55455

JAMES L. DANIEL, Ph.D.
　　　Associate Professor, Pharmacology
　　　Thrombosis Research Center
　　　Temple University Medical School
　　　Philadelphia, PA 19140

OLE-BJØRN TYSNES, M.D., Ph.D.
　　　Associate Professor, Neurology
　　　Preclinical Institute
　　　University of Bergen
　　　Årstadveien 19
　　　N-5009 Bergen
　　　Norway

ADRIE J.M. VERHOEVEN, Ph.D.
　　　Department of Biochemistry
　　　Erasmus University Rotterdam
　　　NO-3000 DR Rotterdam
　　　Holland

GUNDU H.R. RAO, Ph.D.
　　　Professor
　　　Laboratory Medical and Pathology
　　　University of Minnesota
　　　Minneapolis, MN 55455

PREFACE

Platelets, derived from megakaryocytes, are the smallest circulating cellular element in blood. They participate in hemostasis and are responsible for the primary phase of hemostasis. When a blood vessel is damaged, platelets adhere to the subendothelium. Normal adherence of platelets to the subendothelium at high shear forces requires von Willebrand factor (vWf) and the platelet glycoprotein Ib-IX vWf receptor complex. Subendothelial collagen is important because it is a platelet agonist. Platelets that are activated, expose glycoprotein receptors necessary for platelet-to-platelet adherence and form and secrete platelet agonists. Secreted agonists are adenosine diphosphate (ADP) and 5-hydroxytryptamine (5-HT) or serotonin. These platelet agonists are stored within dense granules. The dense granule is so named because of its characteristic ultrastructural appearance and are also known as amine storage organelles, bull's-eye granules, dense bodies, serotonin storage organelles, and very dense granules. Secretion of dense granule constituents is required for normal platelet function. ADP is an important platelet agonist for platelets of every animal species examined to date. Human patients and animals whose platelets cannot secrete dense granule constituents have markedly prolonged bleeding times.

This book examines several aspects of the biology of platelet dense granules. The initial chapter by Dr. James White describes the dense body of human platelets and is followed by a description of the dense granule precursor in megakaryocytes by Drs. Ménard and Meyers. The amine-nucleotide storage complex within the platelet dense granule is discussed by Drs. Holmsen and Ugurbil. The subsequent two chapters by Dr. Daniel and by Drs. Holmsen, Tysnes, and Verhoveven describe the process of granule secretion, focusing on phospholipid metabolism. Platelet activation can be modulated in a positive or negative direction. In some species epinephrine may serve as a gain controller that changes the relationship between the actuating signals and the response. Platelet modulation by epinephrine is discussed in the chapter by Drs. Rao and White. The final chapter by Drs. Meyers and Menard centers on platelet storage pool deficiencies in humans and in animals.

It is hoped that this book will provide a knowledge base that can be used to further advance our understanding of platelet function in health and disease.

Kenneth M. Meyers
Editor

Charles D. Barnes
Editor

Chapter 1

THE DENSE BODIES OF HUMAN PLATELETS

James G. White, M.D.

INTRODUCTION

Blood platelets are secretory cells. Their ability to release substances promoting blood coagulation was recognized by early workers before the turn of the century.[11] However, interest in this aspect of platelet physiology lay fallow for many years due to the lack of methods necessary to study secretory phenomena in such small cells. The work of Kristan Grette changed all that. In his classic monograph[10] on thrombin-catalyzed hemostatic reactions, Grette demonstrated the release of specific chemical substances in parallel from platelets following exposure to thrombin. Discharge of products was not associated with loss of cytoplasmic constituents, indicating that the release reaction was not simply due to thrombin-induced permeabilization. This monumental work established beyond any doubt that blood platelets possessed secretory properties similar to endocrine cells.

Grette's study was limited to evaluation of the action of thrombin on platelets. However, subsequent investigations have shown that nearly all chemical and physical agents capable of triggering irreversible platelet aggregation also stimulate the release reaction.[13] Furthermore, many studies have demonstrated that products made available following platelet activation foster establishment of impervious hemostatic plugs, accelerate conversion of fibrinogen to fibrin, stimulate chemotaxis and cause a proliferation of smooth muscle cells and fibroblasts.[28]

As a result of these studies, interest in platelet secretion was restored and the subject thoroughly evaluated in the 1960s and 1970s.[41] The 1980s saw a shift of research interest away from the platelet release reaction toward more basic aspects of stimulus-activation-contraction-secretion coupling, but the importance of secretion in platelet physiology and pathology remains. This chapter will focus on one form of secretory organelle, the dense body.

DISCOVERY AND CHARACTERIZATION
OF DENSE BODIES

Rand and Ried[19] found that 5-hydroxytryptamine (5-HT, serotonin) was a normal constituent of platelets, and Baker, Blaschko and Born,[2] were able to demonstrate that subcellular particles separated from platelets were rich in this amine as well as in adenosine triphosphate (ATP). Many workers subsequently confirmed the observation of Born's group and added the findings that 5-HT, ATP, and ADP were located either in vacuoles or the granule fraction.[22] The

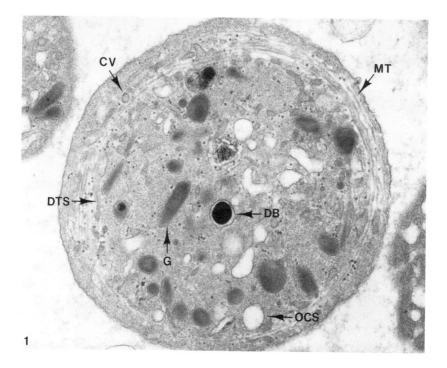

FIGURE 1. Thin section of discoid human platelet. The cell is from a sample of citrate platelet-rich plasma (C-PRP) fixed in glutaraldehyde and osmic acid and embedded in Epon. A circumferential microtubule (MT) supports the discoid form. Granules (G) are randomly dispersed in the cytoplasm. Channels of the open canalicular system (OCS) and dense tubular system (DTS) are also spread throughout the cell. Coated vesicles (CV) and glycogen particles are irregularly dispersed. Electron opaque dense bodies (DB) are less frequent than granules, but their opacity to the electron beam facilitates easy identification (magnification: × 33,000).

subcellular localization of 5-HT was more clearly defined in a number of electron microscopic investigations. Wood[52] reported on the localization of 5-HT at the ultrastructural level, utilizing methods which had been successful in differentiating catecholamine-containing organelles in the adrenal gland. Employing an initial fixation in glutaraldehyde, followed by exposure to potassium dichromate at low or high pH, he was able to separate organelles rich in different amines including 5-HT. In his report, the blood platelet was one of the cells which demonstrated localization of 5-HT in very dense organelles.

Tranzer and his colleagues[26] reported on a slightly different approach to the ultrastructural demonstration of 5-HT in blood platelets. They observed that platelets fixed in glutaraldehyde, followed by osmic acid, contained a variable number of dense osmiophilic bodies, similar to the structures described by Wood, which were not present in platelets exposed to osmic acid alone (Figures 1-3). The number of dense bodies observed in thin sections of platelets from different

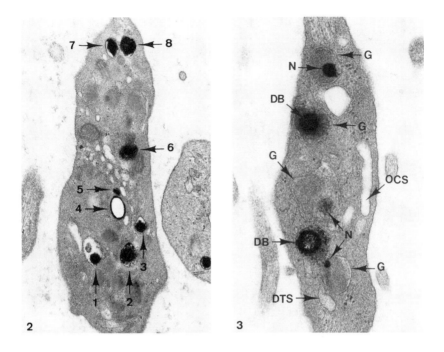

FIGURES 2 and 3. Platelet storage organelles. Granules are the most numerous organelles in platelets, but dense bodies are also present in significant numbers. The cell in Figure 2 contains at least eight dense bodies which are considered the principal storage sites for serotonin (5-HT) and the nonmetabolic pool or adenine nucleotides. Dense bodies vary in size, shape, and content and occasionally appear to contain substance (2) resembling the matrix material of granules. The platelet in Figure 3 also demonstrates the similarity of dense bodies and granules. Granules (G) in this cell contain nucleoids (N) with an opacity similar to that of dense bodies. Dense bodies (DB) are surrounded by material resembling granule matrix. Elements of the open canalicular system (OCS) and dense tubular system (DTS) lie in proximity to both types of secretory organelles (magnifications: Figure 2 × 22,500; Figure 3 × 41,500).

mammalian species appeared to correlate closely with their content of 5-HT. Rabbit platelets rich in 5-HT contained many dense bodies, while human platelets with considerably less 5-HT rarely revealed osmiophilic bodies. Platelets depleted of 5-HT by reserpine or tyramine *in vivo* or *in vitro* were practically devoid of dense bodies, and the content of opaque structures could be restored by incubating depleted platelets with 5-HT. The association of 5-HT with dense bodies was confirmed by ultrastructural autoradiography and by chemical determinations on isolated platelet subcellular organelles, prepared by density gradient centrifugation.

Considerable emphasis was placed on the method of double fixation which resulted in preservation of dense bodies for study in the electron microscope.[8] *In vitro* experiments revealed that combining 5-HT, glutaraldehyde, and osmic acid

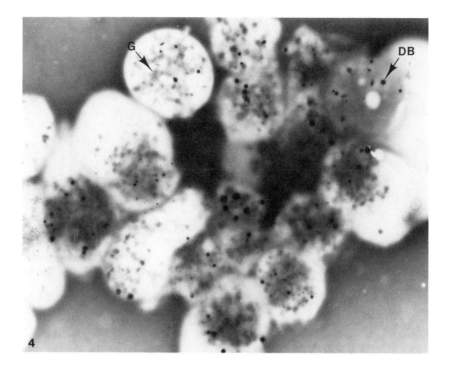

FIGURE 4. The platelets in this illustration were exposed to a small amount of ADP before being mounted on grids and stained with phosphotungstate. Electron dense stain forms a halo around each platelet but does not appear in clear areas of the cytoplasm. Granules (G), however, appear to have been stained by phosphotungstate, though they are not as opaque as dense bodies (DB) (magnification: × 6,200).

resulted in development of a black precipitate. This observation, together with the obvious increased opacity of dense bodies over other platelet organelles, suggested that the 5-HT storage granules were intensely osmiophilic. A similar interpretation was made for their reaction with potassium dichromate at high or low pH in platelets prefixed in glutaraldehyde. It was postulated that glutaraldehyde reacted with 5-HT to form a Schiff base, and interaction of this agent with osmic acid or potassium dichromate would cause precipitation of the metallic-oxidizing fixative. Thus, selective deposition of osmium or dichromate in 5-HT storage organelles was considered to be a specific cytochemical test for 5-HT, as well as the basis for the marked opacity of the dense bodies.

Other investigators questioned both the selective avidity of dense bodies for metallic oxidizing fixatives and osmiophilia as a basis for their electron opacity.[34] Bull,[5] in the course of evaluating negatively stained whole mounts of blood platelets, described the presence of intensely opaque structures localized in the extraneous coat covering the outside surface of the cells. Similar elements had not been observed on the exterior of sectioned platelets fixed by any technique, in

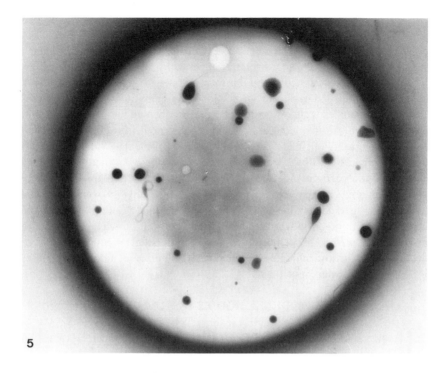

5

FIGURE 5. Platelet from a sample of C-PRP allowed to interact with the grid for 1 min, excess plasma removed and the sample washed on drops of distilled water twice before air drying. This is the most simple and efficient method for analyzing the frequency of platelet dense bodies. The variation in dense body size and density is evident in this example (magnification: × 19,500).

shadow cast preparations or in replicas. This puzzling difference was resolved by demonstrating that the opaque organelles observed by Bull in platelet whole mounts were actually located within the platelet rather than in the exterior coat.[32] Furthermore, the study revealed that the electron opaque organelles in negatively stained platelets, examined by the whole mount technique, were identical to the dense bodies present in the hyaloplasm of sectioned cells fixed in glutaraldehyde and osmium tetroxide (Figure 4).

During this investigation, an extremely important characteristic of platelet dense bodies became apparent. Examination of whole mounted platelets revealed the presence of dense bodies, whether or not they were stained with phosphotungstic acid[33] (Figure 5). All other platelet structures were relatively transparent to electrons or, as in the case of lipids, exhibited a lens effect which concentrated the electron beam. The inherent opacity to electrons suggested that something must be concentrated within the opaque bodies with sufficient mass to trap or otherwise inhibit electron transmission. In biological material, only high concentrations of metallic cations would have the charge density, or deposits of metal salts enough mass, to produce this effect on a beam of electrons.

The observations also raised doubts concerning the specificity of the reaction of dense bodies with osmic acid and other metallic oxidizing fixatives.[33] If the organelles were inherently electron opaque, then their preservation in fixed tissue should not require the enhanced contrast imparted by metallic-oxidizing agents. Experiments were designed to test this thesis. Platelets fixed in glutaraldehyde alone were embedded in plastic and sectioned for study in the electron microscope. Dense bodies were observed in normal numbers in the platelets fixed in aldehyde alone. Although a second fixation in osmic acid or dichromate may enhance the opacity of platelet dense bodies, the experimental evidence suggests that these agents are not essential for their preservation. In fact, the inherent density of the particles appears sufficiently intense to explain the opacity observed in doubly fixed tissue. These findings suggested that osmiophilia had little or nothing to do with the opacity of dense bodies and that the reaction of the organelles with metallic fixatives after aldehyde fixation was not a specific cytochemical method for demonstration of 5-HT in platelets. Subsequent work demonstrated that the inherent electron density of human platelet dense bodies was due to their calcium content.[14]

THE NUMBER OF DENSE BODIES IN HUMAN PLATELETS

Tranzer *et al.*,[26] in the course of their study on the ultrastructural localization of 5-HT in platelets, established a relationship between the concentration of amine and the number of dense bodies in thin sections of the cells. Rabbit platelets, rich in 5-HT, have about 350 of the opaque organelles per 800 μ^2 of platelet surface. Human platelets, low in 5-HT, contained fewer dense bodies than rabbit platelets emptied of amine by prior reserpine treatment of the animals. The rarity of dense bodies in human platelets noted by Tranzer and his colleagues was confirmed in a subsequent study by May *et al.*[15] They observed only 16 dense bodies in thin sections of 1000 normal human platelets. In their study, rabbit platelets were found to contain from 2-10 dense bodies per cell. Thus, the rabbit platelets, with about 20 times more 5-HT than human platelets, appeared to have an average of 375 times more dense bodies per 1000 cells in thin sections than were observed in human platelets. Other workers also suggested that dense bodies occurred rarely in human platelets.[8,15]

An examination of thin sections of glutaraldehyde-osmic acid-fixed platelets in this laboratory revealed a different frequency of 5-HT storage organelles.[32,33] An average of 1-1.4 dense bodies per platelet was found in counts on 100 cells from five normal human donors. Some sectioned platelets had no dense bodies in their matrix. This deficiency, however, was compensated by a significant number of cells containing 4-8 opaque organelles. Evaluation of platelets by the whole mount technique supported the findings made in thin sectioned material. Inherently electron opaque dense bodies were easily counted in the unstained whole mounts (Figure 4). An average of 6.15 dense bodies per platelet was identified with a range of 0-24 per cell (Figure 5).

ORIGIN OF PLATELET DENSE BODIES

Questions concerning the origin of platelet dense bodies still remain. Early work had suggested that formation of the organelles was directly related to the uptake of 5-HT.[26,52] Dense bodies were found only in circulating platelets, never in megakaryocytes. Furthermore, platelets emptied of dense bodies and 5-HT could reform opaque organelles when exposed to 5-HT *in vitro*. Thus, even mature platelets with a full complement of granules and no evidence of granulopoiesis appeared capable of both forming new organelles and sequestering their contents from plasma. Unfortunately, experiments in recent years have not substantiated the early hypotheses.

The appearance of dense bodies as intensely opaque spots, surrounded by a clear zone and a unit membrane, suggested that they formed in the vacuolar system of the platelet (Figure 6). Crawford and his group[7] emphasized the

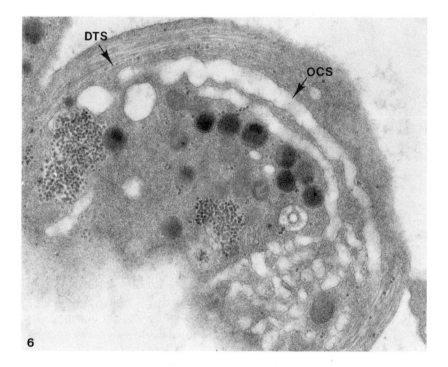

FIGURE 6. Platelet membrane systems. This example is unusual in that a channel of the open canalicular system (OCS), though tortuous, can be followed for a long distance. Elements of the dense tubular system (DTS) are closely associated with the marginal band of microtubules. In one area of cytoplasm components of the two-channel system are interwoven to form a membrane complex[39] (magnification: × 27,500).

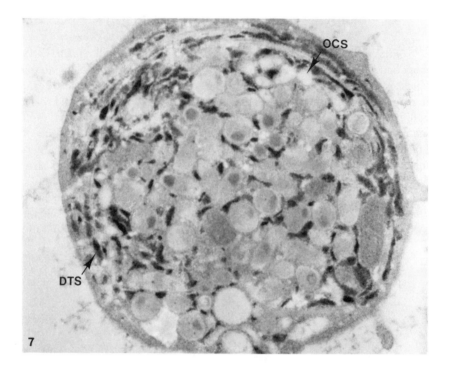

FIGURE 7. Cytochemistry of membrane systems. This platelet is from a sample of platelet-rich plasma incubated for peroxidase activity. Enzyme reaction product is specifically localized to channels of the dense tubular system (DTS) and none is present in the surface-connected canalicular system (OCS) (magnification: × 30,500).

association of 5-HT with vacuoles when they found dilatation of clear channels and increased numbers of vacuoles in platelets from patients with carcinoid syndrome.

There are several possibilities which might serve to explain the 5-HT concentrations in platelet vacuoles. Taken up uniformly over the platelet surface, 5-HT may diffuse through the hyaloplasm and be concentrated in preformed vacuoles.[1] A second possibility would be formation of pinocytotic (coated) vesicles at the surface containing 5-HT, transfer of the vesicles into the hyaloplasm, and fusion of the tiny organelles to make a dense body-sized vacuole[16] (Figure 1).

A third mechanism, similar to the first, could depend on residual elements of endoplasmic reticulum and the Golgi apparatus in platelets.[37] The dense tubular system (DTS) in mature platelets in a residual of endoplasmic reticulum (Figure 7), and a Golgi system can be identified in approximately 1% of circulating cells.[3,39] The 5-HT taken up at the platelet surface may be absorbed in the DTS and transferred via a Golgi apparatus to saccules, vesicles, and ultimately vacuoles.

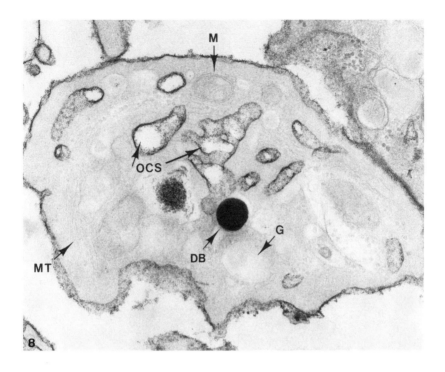

FIGURE 8. Cytochemistry of membrane systems. The platelet was fixed in glutaraldehyde and osmium solution containing lanthanum. Electron dense tracer coats the surface of the cell and lines each channel of the surface-connected canalicular systems (OCS). Microtubules (MT), a mitochondrion (M) and granules (G) are barely visible in this unstained section, but a dense body (DB) is prominent (magnification: × 48,000).

The fourth possibility, akin to the second, would be formation of vacuoles from channels connecting the interior of the platelet to the exterior plasma (Figure 8). The 5-HT extracted from the open channels would be concentrated in the vacuole without entering the hyaloplasm until the vacuole pinched off its connection with the canaliculus. Platelets are known to form coated vesicles (Figure 1) in this same manner.[16]

None of these potential mechanisms for development of dense bodies in platelets has been established. The second and fourth proposals might be eliminated on the basis that the inside surface of vacuoles or vesicles in each case would have to derive from the exterior surface of the platelet. This would require focal sites on the cell wall specialized for the uptake of 5-HT. Also, ADP and ATP would have to reach the outside surface of the cell in order to be concentrated together with 5-HT in vesicles and/or vacuoles. Freeze-fracture would reveal a population of granules in platelets whose enclosing membranes have the same distribution of intramembranous particles on inner and outer faces of their split lipid bilayers as on the cell surface, and the reverse should be true of other

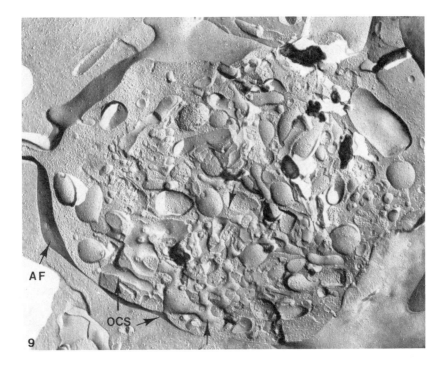

FIGURE 9. The tortuous pathway of the open canalicular system (OCS) and its fenestrated appearance are evident in this replica of a freeze-fractured platelet. The fracture has revealed the outside of the inner lamellae of the lipid bilayer, or A face (AF), in this example. An opening of the OCS on the A face is apparent. The channel extends in all directions, interacting with elements of the DTS and cytoplasmic organelles (magnification: × 32,000).

cytoplasmic storage organelles (Figure 9). However, this has not proven to be the case.[46]

If 5-HT were absorbed evenly over the platelet surface and diffused randomly through the hyaloplasm, it would be difficult to explain why the amine was selectively concentrated in a few vacuoles. Selection of certain vacuoles for dense body formation would imply that the preformed organelles were specially adapted for this purpose, whereas other clear elements in the hyaloplasm were not. The fourth possibility escapes this criticism by suggesting that the route followed by 5-HT in platelet hyaloplasm is similar to that followed by chemical constituents during granulopoiesis. As a result, new Golgi vesicles and vacuoles would be formed under the influence of 5-HT uptake and their condensation would cause dense body formation. Although such a mechanism might obviate the problems of the first possibility proposed, at present there is very little evidence to support it.

The Golgi apparatus in circulating platelets is a remnant of the organelle in parent megakaryocytes and probably lacks function in the end stage platelet.

Vesicles derived from it have not been recognized as a distinct population in the circulating cell. The vesicles and vacuoles in platelets are almost exclusively part of the tortuous open canalicular system (OCS) (Figure 8) based on studies with electron dense tracers.[36,40] Thus, very few free vesicles would be available in mature platelets for participation in the process of dense body formation.

One central error in all four of the proposed mechanisms, however, may be the assumption that dense bodies are formed in vacuoles. Studies in our laboratory suggested an alternative mechanism for dense body formation. Examination of many electron photomicrographs suggested various stages in the transformation of platelet granules to dense bodies[31] (Figures 10, 11). The process appeared to involve the more dense of the two zones in the matrix of the granule. An intensely opaque concentric body replaced the nucleoid in the α granule. Since matrix material may continue to surround the dense spot or may diminish in amount, internal lysis is suggested. Such a transformation could explain why dense bodies usually have a clear space separating the opaque core from the enclosing membrane. However, recent ultrastructural studies employing analytical techniques have suggested that the elemental composition of opaque nucleoids in α granules differs from that of dense bodies. Thus, electron-dense nucleoids of α granules are probably unrelated to classical dense bodies. Precisely what they are and how they develop in α granules remains uncertain.

Another possible error inherent in the above argument is that all dense bodies may have developed in at least a putative form before the platelet leaves the megakaryocyte. Many years ago we demonstrated that dense bodies were present in megakaryocytes from normal human bone marrows.[37] Tranzer *et al.*[27] confirmed this observation in guinea pigs and showed that intraperitoneal injection of 5-HT resulted in a marked increase in density and number of dense bodies present in megakaryocytes developing *in situ*.

If dense bodies are present in megakaryocytes, some mechanism must trigger their development in the parent cell. Recent studies have suggested that such a mechanism does exist. Employing the uranaffin reaction introduced by Richards and Da Prada,[21] Daimon and Gotoh[9] confirmed the presence of dense bodies in megakaryocytes. The uranaffin-stained particles appeared to arise in vesicles leaving the face of the Golgi zone opposite to that involved in α granule genesis. Unfortunately, flattened saccules of the Golgi apparatus were not stained by uranium. As a result, it remains uncertain whether the vesicles were present in the vicinity of Golgi zones by chance or if saccules of the apparatus are the definitive source of membranes for dense bodies.

Recently we have been impressed that elements of the DTS may be involved in dense body formation. Small, irregular, electron-dense nucleoids are often visible in thin sections of normal platelets in what resemble elements of the DTS. They are also apparent in thin sections and whole mounts of a few patients with Hermansky-Pudlak syndrome (HPS) (Figures 12, 13),[50] whose platelets are nearly devoid of dense bodies. A more impressive example of possible DTS involvement was found in a family with giant platelet dense bodies (Figure 14).

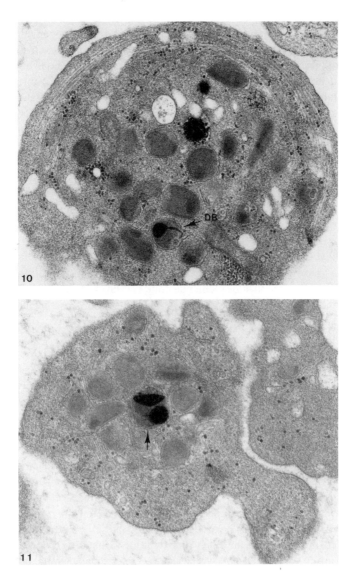

FIGURES 10 and 11. Serotonin (5-HT) storage organelles. Previous studies have suggested that most of the 5-HT storage organelles in human platelets originate by transformation of granules. The more opaque of the two zones present in each granule appears to be the focal point of transformation. One granule (indicated by arrow) in the platelet shown in Figure 10 has developed an electron-dense spot similar in opacity to the dense body in the cytoplasm. The tail-like extension is frequently observed attached to 5-HT storage organelles. Figure 11 shows a platelet from a sample of C-PRP fixed during the first wave of aggregation stimulated by ADP. Among the centrally grouped organelles is a granule (indicated by arrow) containing two opaque zones identical in morphology to dense bodies (magnifications: Figure 10 × 27,000; Figure 11 × 41,000).

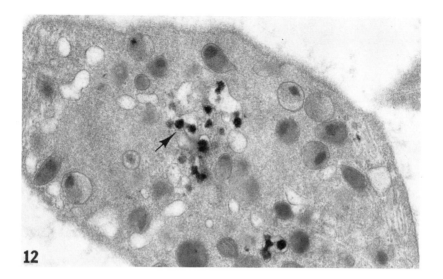

FIGURE 12. Thin section of a platelet from a patient with Hermansky-Pudlak Syndrome (HPS). Platelets from most patients with HPS are devoid of dense bodies, but some contain dense nucleoids (indicated by arrow) which may be precursors of the 5-HT rich organelles in normal platelets (magnification: × 21,000).

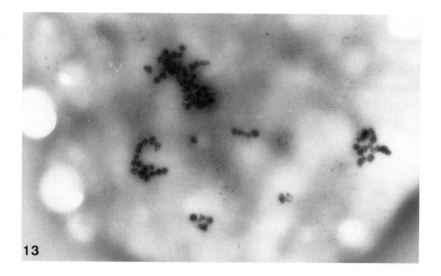

FIGURE 13. Whole mount of a platelet from the same patient shown in the previous illustration. Nucleoids in his cells are inherently electron opaque. This patient differs from most patients with HPS who lack dense bodies and nucleoids (magnification: × 16,500).

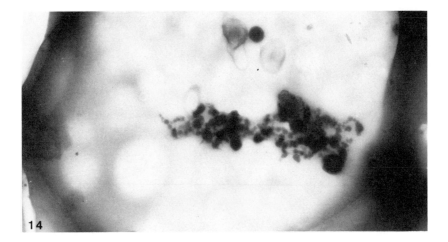

FIGURE 14. Whole mount of platelet from one of our patients with a giant dense body disorder. Large numbers of nucleoids are clustered together in these cells and may represent precursors of dense bodies (magnification: × 16,500).

All stages in formation of the large organelles from nucleoids, present in very large numbers, are evident in their platelets. The DTS is known to sequester calcium and make the cation available during physiologic activities.[24] Therefore, DTS channels contain proteins with divalent cation-binding capability. Uptake and concentration of calcium, along with nucleotides and 5-HT may be steps in formation of the opaque storage organelles.

The precise origin of the dense bodies in human platelets remains uncertain. They can form in megakaryocytes and, after leaving the parent cell, in circulating platelets. Evidence has suggested that they arise from the Golgi zone in megakaryocytes or vesicles, α granules and channels of the DTS in circulating cells. Further effort will be required before their specific origin can be unequivocally defined.

SECRETION OF PLATELET DENSE BODIES

Platelets respond to a wide variety of chemical, particulate, and other stimuli in a characteristic manner.[43] The cells quickly lose their discoid shape, become relatively spherical, and extend long, spiky pseudopods, as well as bulky surface protrusions. Changes in the surface contour are accompanied by movement of randomly dispersed internal organelles toward platelet centers. The transfer of granules and dense bodies is associated with constriction of the circumferential band of microtubules and contraction of the filamentous matrix of the sol-gel zone. In fully activated platelets, the organelles are clumped together in the central region of the cell and enclosed by a tight-fitting ring of microtubules and microfilaments. As contents of the secretory organelles are extruded, the web of tubules and filaments continues to contract until only a mass of contractile gel

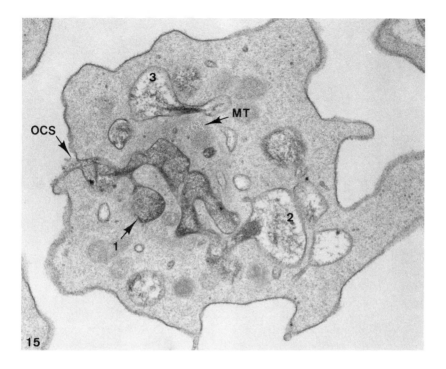

FIGURE 15. EDTA washed platelet exposed to 3 U/ml of thrombin for five min. Stained (1) and swollen (2) granules are directly connected to the channel filled with material reacting with tannic acid-osmium. Other swollen granules, (3), are connected to different channels of the OCS and pass out of the plane of section. Microtubules (MT) are constricted into the central mass of contractile gel (magnification: × 33,000).

remains in the central zone. The constricted ring of tubules appears to fracture, and individual elements dissolve or are dispersed, probably to pseudopods.

The degree to which platelets respond to a particular stimulus depends on the nature of the agent, its concentration, the rate of stirring, and the sensitivity of the cell population.[36] Also, no two platelets appear to be at exactly the same stage of transformation at the same time. Thus, the physical changes can occur before, during, and after aggregation. Although some platelets appear to have completed their response, others at the same moment remain unaltered. This heterogeneity, providing a bell-shaped curve of response, is typical for the reactivity of biologic systems.

Alpha granule secretion from human and bovine platelets has been addressed in several publications.[38,44] The process in human platelets involves fusion of granules with channels of the OCS and discharge of products through channels to the outside (Figures 15-20). Bovine cells lack the OCS present in human and most animal platelets. Alpha granules move to the surface of activated bovine platelets, fuse with the membrane and discharge directly to the exterior.[45]

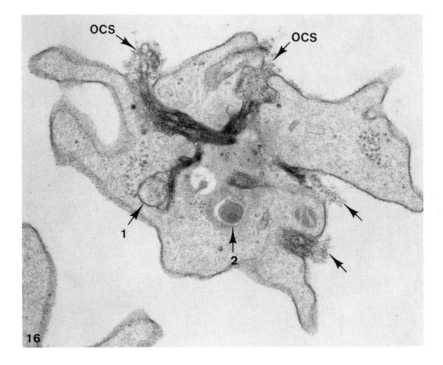

FIGURE 16. Platelet from sample of EDTA washed platelets exposed to thrombin at 3 U/ml for one min. Electron-dense material is in the process of extrusion from four different sites (indicated by arrows) on the surface of this cell. One of the channels inside the cell communicates with two different openings of the OCS and with a granule (1). Other granules (2) are not stained by tannic acid-osmium (magnification: × 33,000).

Dense body secretion was investigated in early studies.[35,38] Experiments showed that polylysine and polybrene stabilized contents of dense bodies and permitted recognition during their extrusion through channels of the OCS to platelet exteriors (Figures 17-20). Other workers have suggested that dense bodies are secreted in a manner different from α granules. For example, some believe that dense bodies fuse with the surface membrane and discharge to the outside,[23] while others suggest that the organelles leak their contents to the cytoplasm. The latter mechanism is unlikely because contents of dense bodies, including calcium, pyrophosphate, adenine nucleotides and 5-HT are known to be secreted to the platelet exterior.

The differences in opinion are related to the problems of clearly identifying dense body secretion. Their contents dissolve immediately on contact with surrounding media or plasma, and dense body membranes are indistinguishable from those enclosing other organelles. Attempts to use the uranaffin reaction[21] to selectively stain dense body membranes during secretion have not been successful.

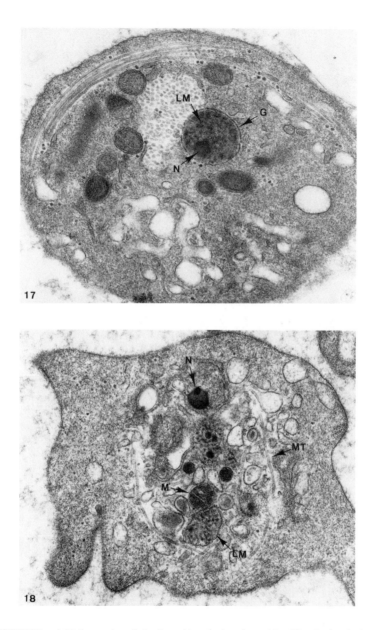

FIGURES 17 and 18. Interaction of platelets with cationic polypeptides. The platelets in the two illustrations were incubated with polylysine and polybrene respectively. Both agents produce identical effects. They are taken up and deposited in platelet secretory organelles. The light matrix (LM) of the platelet granules is polymerized into a lattice-like structure. Granule nucleoids (N) become dense and fragmented. After prolonged incubation, the cells lose their discoid form and develop internal transformation (magnifications: Figure 17 × 36,000; Figure 18 × 41,000).

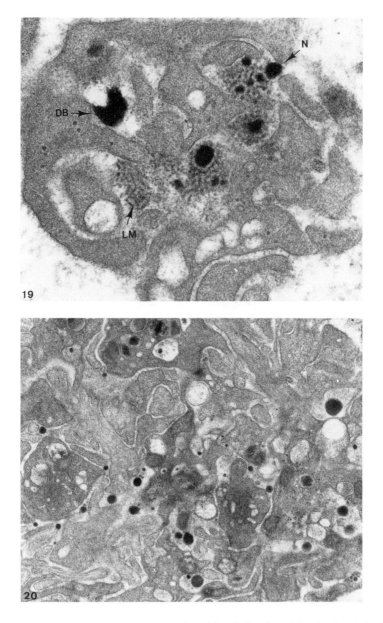

FIGURES 19 and 20. After prolonged incubation with cationic polypeptides, the altered platelets extrude their storage organelles. A portion of a granule consisting of the nucleoid and polymerized light matrix is evident at the periphery of the platelet in Figure 19. Figure 20 identifies the secretory pathway in aggregated platelets. The platelet sample was fixed 5 min after exposure to collagen on the platelet aggregometer. Polylysine was added to the sample just prior to collagen and the stirring rod. Storage organelles in all stages of extrusion are easily identified (magnifications: Figure 19 × 41,000; Figure 20 × 13,000).

Dense body release from platelets appears similar to α granule secretion; however, there are some differences. A definitive method for identifying membranes of dense bodies may soon be at hand, and remaining questions can be resolved.

INHERITED DISORDERS OF DENSE BODIES

Hermansky-Pudlak Syndrome (HPS)

The HPS is a recessively inherited autosomal disease in which the triad of tyrosinase-positive oculocutaneous albinism, accumulation of ceroid-like material in reticuloendothelial cells of bone marrow and other tissues, and a hemorrhagic diathesis owing to defective platelets are constantly associated.[51] Well over 100 cases of this syndrome have been reported in the world literature. In previous reviews of ultrastructural defects in congenital disorders of platelet function,[40,47,48] it was suggested that HPS was the first disorder in which an abnormality detectable in the electron microscope could be correlated directly with a specific biochemical deficiency, i.e., impaired platelet function *in vitro* and clinical bleeding problems in patients. The population of electron-dense bodies in HPS platelets is greatly reduced and, in some cases, absent (Figures 21, 22). Yet, biochemical analyses revealed the same sequential changes as normal platelets when HPS cells were stimulated by aggregating agents. These included shape change, internal transformation, and molding of cell surfaces together in tightly packed small aggregates. Owing to the marked deficiency in ADP, the amount of nucleotide secreted by minimally activated HPS cells was insufficient either to bring uninvolved platelets into large aggregates or to sustain the platelet-platelet association long enough to establish irreversible aggregation. As a result, HPS platelets do not develop second waves of aggregation when exposed to concentrations of ADP, epinephrine, and thrombin, which cause biphasic, irreversible aggregation of normal cells on the aggregometer. However, they will form irreversible aggregates if exposed to a high concentration of exogenous ADP. In addition, they can form irreversible aggregates, in some instances, even when stirred with epinephrine alone.[48]

The latter observation suggested that HPS platelets might possess a mechanism capable of compensating for the storage pool deficiency (SPD). Such a mechanism, called membrane modulation, has been described.[20] It was first observed in dog platelets, which are insensitive to epinephrine, and, in two-thirds of the animals, unreactive to arachidonic acid. However, when dog platelets were exposed to small amounts of epinephrine before stirring with arachidonic acid, they developed irreversible aggregation.

A similar mechanism mediated by α-adrenergic receptors is present in normal human platelets. Aspirin inhibits the platelet response to arachidonate. Exposure to a small amount of epinephrine, however, restores the sensitivity of aspirin-treated cells and permits development of irreversible aggregation when stirred with arachidonate. Reversal of the refractory state was not associated with

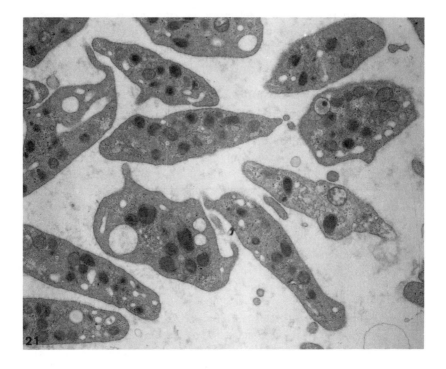

FIGURE 21. Thin section of platelets from a patient with HPS. Dense bodies are absent from these cells (magnification: × 16,500).

restoration of prostaglandin synthesis or secretion. Membrane modulation also was found to restore the sensitivity of platelets made refractory by prior aggregation and disaggregation or exposure to antibiotics. Dr. Rao has discussed this mechanism in more detail in Chapter 6 of this volume.

The possibility that a similar mechanism operates in HPS platelets was evaluated. Platelets from a patient with HPS were treated with aspirin to block prostaglandin synthesis. His platelets were unable to convert arachidonic acid into thromboxane A_2 or to secrete ADP and other constituents of the release reaction. Exposure to small quantities of epinephrine, unable to cause aggregation of the aspirin-treated normal or HPS platelets, restored the sensitivity of the refractory cells. Despite inability to secrete or form thromboxane, the refractory HPS platelets developed irreversible aggregation when stirred with arachidonate after prior treatment with epinephrine. A similar alteration in sensitivity to other aggregating agents was also noted. Thus, HPS platelets possess the mechanism of membrane modulation, which may compensate to some degree for SPD *in vivo*.

The nature of the HPS-associated abnormalities, including tyrosinase-positive albinism, ceroid-lipofuscin storage, pulmonary infiltration, and colitis, and their relationships to the platelet SPD remain unknown. Histochemical

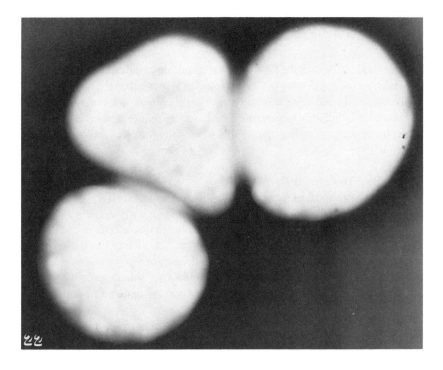

FIGURE 22. Whole mount of HPS cells. Dense bodies and nucleoids are absent from the HPS platelets (magnification: × 13,000).

studies have shown similarities between ceroid-lipofuscin in HPS macrophages and aging pigment. However, the amount of the material produced in the HPS and its wide distribution in the reticulo-endothelial system suggest a relationship to other sea blue histiocyte syndromes.[25] The lipoprotein complexes making up the stored material in HPS have as yet resisted biochemical analysis.

Clinical problems of HPS patients vary widely in severity. The original patients may have died from hemorrhagic problems. Others have had severe life-threatening bleeding episodes after dental extraction or gastrointestinal hemorrhage. Yet, logic would suggest that absence or marked deficiency of the platelet storage pool should not, in and of itself, be a major cause of bleeding. Ingestion of aspirin has nearly the same effect as the inherited disorder, because it blocks the platelet-release reaction. Although the influence of aspirin can be overcome by thrombin, the defect in HPS platelets is compensated by the mechanism of membrane modulation discussed above. Aspirin causes few, if any, severe bleeding symptoms in normal individuals. Similarly, most patients with HPS have very mild hemorrhagic problems. The major difficulty is related to their gastrointestinal or pulmonary disease and its associated complications. Yet, the history of the original patients and others reported in the literature would suggest that a significant number do have serious bleeding. The basis for the difference between HPS patients with mild or severe hemorrhagic problems has not been clearly defined.

The cause of SPD in the HPS and its relationship to the other components of the syndrome remain unknown. Current research is focused on the possibility that membranes enclosing platelet-dense bodies are abnormal or not produced in the megakaryocyte. Inability to generate membranes or failure of the membranes to concentrate cations may link the SPD to pseudoalbinism and ceroid lipofuscin accumulation.

Storage Pool Disease (SP*D*)

Platelet SP*D* had been reported to be less frequent than HPS.[28,29] However, it may be more common. Because bleeding symptoms are mild and the pseudoalbinism and ceroid-lipofuscin accumulation characteristic of HPS are absent, individuals with SP*D* may not come to the attention of physicians. Also, the degree of adenine nucleotide and 5-HT deficiency in SP*D* platelets is variable and usually not as severe as in the HPS, resulting in further moderation of the disease. As a result, many patients with SP*D* may go undetected during their lifetimes.

Weiss and colleagues[30] have provided an excellent analysis of this condition. Of 18 patients with various granule disorders, four were found to have dense-body deficiency without other clinical features of HPS. In at least one family, the disorder appeared to be inherited as an autosomal dominant. The hemorrhagic symptoms in SP*D* patients were generally mild, and they lacked the severe gastrointestinal bleeding seen in some patients with HPS. Depletion of dense-body contents and electron opaque organelles was less in SP*D* platelets compared with individuals with HPS. Weiss had noted that 5-HT levels in SP*D* platelets were reduced in proportion to the reduction in platelet ATP.

The ability of SP*D* platelets to absorb ^{14}C-5-HT may be more compromised than observed in HPS cells. The SP*D* platelets take up 5-HT initially at the same rate as normal cells but are quickly saturated. The 5-HT accumulated by normal platelets is retained, but the absorbed 5-HT in SP*D* platelets is discharged steadily in the form of metabolites. When SP*D* platelets prelabeled with ^{14}C-5-HT are stimulated by aggregating agents, they release the ^{14}C-5-HT more slowly than do normal cells, suggesting that the release reaction may be defective in SP*D*. SP*D* platelets were also deficient in their ability to synthesize intermediates of prostaglandin biosynthesis. After stimulation by collagen, SP*D* platelets produced less than 20% of the prostaglandin E_2 and prostaglandin $F_2\alpha$ synthesized by normal cells.

The ultrastructural defect in SP*D* platelets appears essentially identical to that observed in HPS. Morphologic characteristics are similar to normal platelets, except for the profound reduction in the number of dense bodies. The decrease in dense bodies correlates not only with the deficiency in 5-HT and adenine nucleotides, but also with the impaired cellular response to aggregating agents and with the clinical symptoms of the patients. Thus, SPD is the second disorder in which impaired platelet function can be associated directly with an ultrastructural defect in the cells. However, the normal pigmentation of individuals with this disorder and the absence of an unusual accumulation of ceroid or lipofuscin in macrophages suggest that the cause is basically different than that responsible for platelet SPD in HPS. Further studies will be required to clarify this difference.

Chediak-Higashi Syndrome (CHS)

The CHS[6,12] is a rare, autosomally inherited disorder characterized clinically by photophobia, nystagmus, pseudoalbinism, marked susceptibility to infection, hepatosplenomegaly, lymphadenopathy, and early death, frequently owing to development of an accelerated phase resembling a viral-associated hemophagocytic syndrome. Laboratory diagnosis is based on the presence of giant organelles in

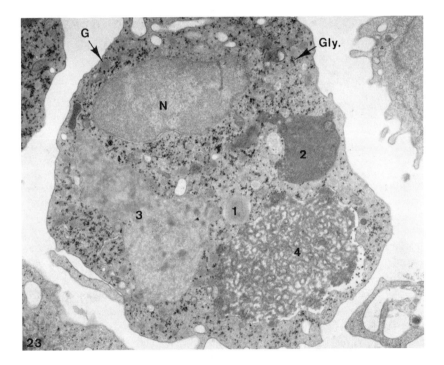

FIGURE 23. CHS neutrophil. Thin section of a PMN from a sample of leukocytes fixed in glutaraldehyde and osmic acid containing potassium ferrocyanide. The cell contains a portion of the nucleus (N), numerous glycogen particles (Gly), an abundance of normal-sized granules (G), and several giant organelles. One (1) of the CHS organelles is small compared to the others and spherical in form. Another (2) is irregular and relatively electron dense. A third (3) is very large and fills a considerable area of the cytoplasm. The fourth (4) is almost as large but radically different in substructure. Its matrix appears composed of a complex array of membranes not apparent in any of the other giant granules in this cell (magnification: × 13,000).

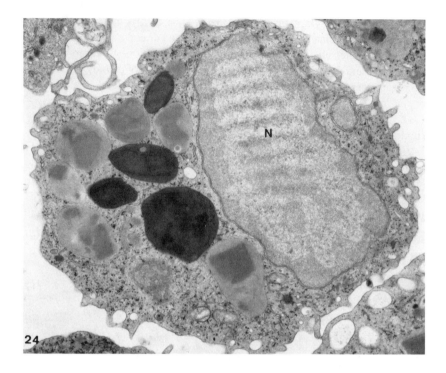

FIGURE 24. CHS eosinophil. The giant lysosomes of CHS eosinophils are huge, but resemble normal size eosinophil granules in most respects. Some are irregular in shape, but most are round or oval. Surface contours are smooth and the substructure is like that of granules in normal cells, except for the presence of several crystals rather than one. Giant organelles filled with debris of the type common in CHS neutrophils (lysosome 4 in Figure 1) are not found in patient eosinophils (magnification: × 13,000).

nearly all leukocytes on Wright-stained peripheral blood smears. The massive granules have been found in neutrophils, eosinophils, lymphocytes and monocytes from blood and in their bone marrow precursors (Figures 23, 24).

Despite thrombocytopenia, which develops during the accelerated phase of CHS, and an early report describing two patients with markedly decreased platelet 5-HT,[17] blood platelets were not considered a major problem in this disease. However, several studies have shown that platelets express the genetic fault of the disorder. Platelets from patients with CHS are biochemically, physiologically and functionally abnormal. The defect has been related to a profound deficiency in the storage pool of adenine nucleotides and 5-HT and a marked reduction in platelet dense bodies.[4] Elevated levels of cyclic 3'5'-adenosine monophosphate (cAMP) were noted in platelets from one infant with CHS. However, the level of cAMP was corrected to normal by treatment with ascorbate without apparent improvement in platelet function. Thus, the platelet

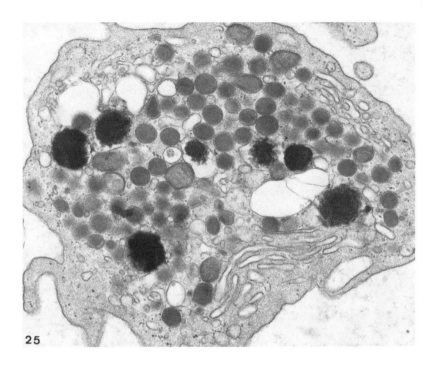

FIGURE 25. Thin section of platelet from one of our patients with a giant dense body disorder. Four of the seven dense bodies in the cell are several times larger than α granules (magnification: × 25,000).

appears to be involved in the expression of the CHS along with other blood cells containing giant cytoplasmic granules.

Involvement of platelets in the CHS appears variable. A patient with characteristic clinical and laboratory features of the disease has been followed in this laboratory for 15 years. His platelets are functionally, biochemically, and morphologically close to normal. However, the author has studied several other patients with CHS who have SPD and their platelets are nearly devoid of dense bodies. Our patient has about 10-20% of the normal number of dense bodies per platelet.

In addition to SPD, platelets from patients with CHS also have been found to contain the giant granule anomaly.[42] Giant granules of a type not seen in normal platelets nor in other platelet disorders were found in thin sections of platelets from our CHS patient in a ratio of about 1 per 1000 cells. Parmley and associates[18] have confirmed this observation in another patient and have shown that the giant granules in CHS platelets are acid phosphatase-positive. A third patient with CHS has been evaluated recently and found to have similar giant platelet granules. The relationship of the giant granule anomaly to the SPD of CHS platelets has yet to be defined.

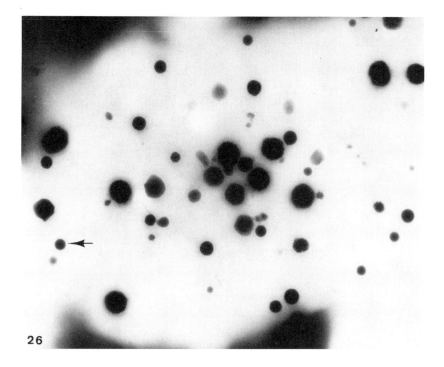

26

FIGURE 26. Whole mount of platelets from one of our patients with a giant dense body disorder. The arrow indicates a normal-sized dense body. Opaque organelles are markedly increased in number and size in this cell (magnification: × 15,000).

Other Dense Body Disorders

Weiss, *et al.*[30] have described a condition in which dense bodies and α granules are profoundly decreased. Unfortunately, only one patient with this platelet problem has been reported. As a result, it is not certain whether her bleeding problems are due to decreased dense bodies, α granules or some other platelet deficiency.

We have described a mother and child whose platelets contain increased numbers of dense bodies, many that are very large[49] (Figures 25, 26). Efforts to define the basis for their abnormalities are not complete. However, preliminary studies suggest the presence of two populations of dense bodies in their platelets. One may be of lysosomal origin because giant organelles in both maternal and child platelets are peroxidase positive. The other population may arise in a normal manner because levels of 5-HT and adenine nucleotides are normal in both patients. A defect in calcium flux is also present in their cells, but its discussion is beyond the purview of this report.

SUMMARY

The material reviewed in this chapter has attempted to bring the reader a current view of human platelet dense bodies. It is far from complete. Knowledge required to answer questions concerning the origin, development, frequency, function, and pathology of the opaque structures is accumulating rapidly. By the time this subject is reviewed again, most of the remaining questions will have been answered.

List of Abbreviations
5-HT = serotonin (5-hydroxytryptamine)
ATP = adenosine triphosphate
cAMP = cyclic 3'5'-adenosine monophosphate
CHS = Chediak-Higashi Syndrome
C-PRP = citrate platelet-rich plasma
DTS = dense tubular system
HPS = Hermansky-Pudlak Syndrome
OCS = open canalicular or surface-connected canalicular system
SPD = storage pool deficiency
SP*D* = storage pool disease

REFERENCES

1. **Bak, I.J., Hassler, R., May, B., and Westerman, E.,** Morphological and biochemical studies on the storage of serotonin and histamine in blood platelets of the rabbit, *Life Sci.,* 6, 1133-1146, 1967.
2. **Baker, R.V., Blaschko, H., and Born, G.V.R.,** The isolation from blood platelets of particles containing 5-hydroxytryptamine and adenosine triphosphate, *J. Physiol. (Lond.),* 149, 55-61, 1959.
3. **Behnke, O.,** The morphology of blood platelet membrane systems, *Ser. Hematol.,* 3, 3-16, 1970.
4. **Boxer, G.J., Holmsen, H., Robkin, L., Bang, N.U., Box, L.A., and Baehner, R.L.,** Abnormal platelet function in Chediak-Higashi syndrome, *Br. J. Haematol.,* 35, 521-533, 1977.
5. **Bull, B.S.,** The ultrastructure of negatively stained platelets: Some physiological interpretations, *Blood,* 28, 901-912, 1966.
6. **Chediak, M.,** Nouvelle anotamie leukocytaire de caractere constitutional et familia, *Rev. Hematol. Paris,* 7, 362-372, 1952.
7. **Crawford, N., Sutton, M., and Horsfield, G.I.,** Platelets in the carcinoid syndrome: A chemical and ultrastructural investigation, *Br. J. Haematol.,* 73, 181-195, 1967.
8. **Da Prada, M., Pletscher, A., Tranzer, J.P., and Knuchel, H.,** Subcellular localization of 5-hydroxytryptamine in blood platelets, *Nature,* 216, 1315-1317, 1967.
9. **Daimon, T., and Gotoh, Y.,** Cytochemical evidence of the origin of the dense tubular system in the mouse platelet, *Histochemistry,* 76, 189-196, 1982.
10. **Grette, K.,** Studies on the mechanism of thrombin-catalyzed hemostatic reaction in blood platelets, *Acta Physiol. Scand. Supp.,* 56(195), 1-93, 1962.
11. **Hayem, G.,** Sur le mechanism de l'arret des hemorrhagies. C.r. Hebd. Seanc., *Acad. Sci. Paris,* 95, 18-36, 1882.
12. **Higashi, O.,** Congenital gigantism of peroxidase granules: The first case ever reported of qualitative abnormality of peroxidase, *Tohoku J. Exp. Med.,* 59, 315-321, 1954.

13. **Holmsen, H., Day, H.J., and Stormorken, H.,** The blood platelet release reaction, *Scand. J. Haematol. Suppl.,* 8, 3-26, 1969.

14. **Martin, J.H., Carson, F.L., and Race, G.J.,** Calcium containing platelet granules, *J. Cell Biol.,* 60, 775-777, 1974.

15. **May, B., Bak, I.J., Bohle, E., and Hassler, R.,** Electron microscopial and biochemical studies on the serotonin granules in carcinoid syndrome, *Life Sci.,* 7, 785-800, 1968.

16. **Morgenstern, E.,** Coated membranes in human platelets, *Eur. J. Cell Biol.,* 26, 315-318, 1982.

17. **Page, A.R., Berendes, H., Warner, J., and Good, R.A.,** The Chediak-Higashi syndrome, *Blood,* 20, 330-338, 1962.

18. **Parmley, R.T., Poon, M.C., Crist, W.M., and Molluk A.,** Giant platelet granules in a child with the Chediak-Higashi syndrome, *Am. J. Hematol.,* 6, 51-60, 1979.

19. **Rand, M., and Ried, G.,** Source of serotonin in serum, *Nature,* 168, 385-386, 1951.

20. **Rao, G.H.R., Reddy, K.R., and White, J.G.,** Modification of human platelet response to sodium arachidonate by membrane modulation, *Prost. Med.,* 6, 75-90, 1981.

21. **Richards, J.G., and Da Prada, M.,** Uranaffin reaction: A new cytochemical technique for the localization of adenine nucleotides in organelles storing biogenic amines, *J. Histochem. Cytochem.,* 25, 1322-1336, 1977.

22. **Siegel, A., and Luscher, E.F.,** Non-identity of the granules of human blood platelets with typical lysosomes, *Nature,* 215, 745-746, 1967.

23. **Skaer, R.J.,** Platelet Degranulation, in *Platelets in Biology and Pathology, Vol. 2,* Gordon, J., Ed., Elsevier/North-Holland Biomedical Press, Amsterdam, 1981, 321-348.

24. **Statland, B., Heagan, B., and White, J.G.,** Uptake of calcium by platelet relaxing factor, *Nature,* 223, 521-523, 1969.

25. **Takahashi, A., and Yokoyama, T.,** Hermansky-Pudlak syndrome with special reference to lysosomal dysfunction, *Virchows Arch. [Path. Anat.],* 402, 247-258, 1984.

26. **Tranzer, J.P., Da Prada, M., and Pletscher, A.,** Letter to the editor: Ultrastructural localization of 5-hydroxytryptamine in blood platelets, *Nature,* 211, 1574-1575, 1966.

27. **Tranzer, J.P., Da Prada, M., and Pletscher, A.,** Storage of 5-hydroxytryptamine in megakaryocytes, *J. Cell Biol.,* 52, 191-197, 1972.

28. **Weiss, H.J.,** Abnormalities in platelet function due to defects in the release reaction, *Ann. N.Y. Acad. Sci.,* 201, 161-173, 1972.

29. **Weiss, H.J.,** Platelet physiology and abnormalities of platelet function, *N. Engl. J. Med.,* 293, 531-580, 1975.

30. **Weiss, H.J., Witte, L.D., Kaplan, K.L., Lages, B.A., Chernoff, A., Nossel, H.L., Goodman, D.S., and Baumgartner, H.R.,** Heterogeneity in storage pool deficiency: Studies on granule-bound substances in 18 patients including variants deficient in alpha granules, platelet factor-4, beta-thromboglobulin and platelet-derived growth factor, *Blood,* 54, 1296-1308, 1979.

31. **White, J.G.,** The dense bodies of human platelets: Origin of serotonin particles from platelet granules, *Am. J. Pathol.,* 53, 791-808, 1968.

32. **White, J.G.,** The origin of dense bodies in the surface coat of negatively stained platelets, *Scand. J. Haematol.,* 5, 371-382, 1968.

33. **White, J.G.,** The dense bodies of human platelets: Inherent electron opacity of serotonin storage particles, *Blood,* 33, 598-606, 1969.

34. **White, J.G.,** Origin and function of platelet dense bodies, *Series Haematologica,* 3, 17-46, 1970.

35. **White, J.G.,** A search for the platelet secretory pathway using electron dense tracers, *Am. J. Pathol.,* 58, 31-49, 1970.

36. **White, J.G.,** Platelet morphology, in *The Circulating Platelet,* Johnson, S.A., Ed., Academic Press, New York, 1971, 45-121.

37. **White, J.G.,** Serotonin storage organelles in human megakaryocytes, *Am. J. Pathol.,* 63, 403-410, 1971.

38. **White, J.G.,** Exocytosis of secretory organelles from blood platelets incubated with cationic polypeptides, *Am. J. Pathol.,* 69, 441-453, 1972.

39. **White, J.G.,** Interaction of membrane systems in blood platelets, *Am. J. Pathol.,* 66, 295-312, 1972.

40. **White, J.G.,** Ultrastructural defects in congenital disorders of platelet function, *Ann. N.Y. Acad. Sci.,* 201, 205-233, 1972.

41. **White, J.G.,** Electron microscopic studies of platelet secretion, *Prog. Hemost. Thromb.,* 2, 49-98, 1974.

42. **White, J.G.,** Platelet microtubules and giant granules in the Chediak-Higashi syndrome, *Am. J. Med. Tech.,* 44, 273-278, 1978.

43. **White, J.G.,** The morphology of platelet function, in *Methods in Hematology, Series 8L, Measurements of Platelet Function,* Harker, L.A., and Zimmerman, T.S., Eds., Churchill-Livingstone, New York, 1983, 1-25.

44. **White, J.G.,** Further studies of the secretory pathway in thrombin-stimulated human platelets, *Blood,* 69, 1196-1203, 1987.

45. **White, J.G.,** The secretory pathway of bovine platelets, *Blood,* 69, 878-885, 1987.

46. **White, J.G., and Conrad, W.J.,** The fine structure of freeze-fractured blood platelets, *Am. J. Pathol.,* 70, 45-56, 1973.

47. **White, J.G., and Gerrard, J.M.,** Ultrastructural features of abnormal blood platelets, *Am. J. Pathol.,* 83, 590-632, 1976.

48. **White, J.G., and Gerrard, J.M.,** The ultrastructure of defective human platelets, *Mol. Cell. Biochem.,* 21, 109-128, 1978.

49. **White, J.G., Smithson, W.A., McCaffrey, L.A., and Rao, G.H.R.,** Platelet hypercalcemia and giant dense bodies: A new familial disorder, *Blood,* 68, 312a, 1986.

50. **White, J.G., and Witkop C.J., Jr.,** Effect of normal and aspirin platelets on defective secondary aggregation in the Hermansky-Pudlak syndrome: A test for storage pool deficient platelets, *Am. J. Pathol.,* 68, 57-66, 1972.

51. **Witkop, C.J., Jr., Hill, C.W., Desnick, S.J., Thies, J.F., Thorn, H.L., Jenkins, M., and White, J.G.,** Ophthalmologic, biochemical, platelet and ultrastructural defects in various types of oculocutaneous albinism, *J. Invest. Dermatol.,* 60, 443-456, 1973.

52. **Wood, J.G.,** Electron microscopic localization of 5-hydroxytryptamine (5-HT), *Texas Rep. Biol. Med.,* 23, 828-837, 1965.

Chapter 2

ULTRASTRUCTURE OF DENSE GRANULE PRECURSOR DEVELOPMENT IN MEGAKARYOCYTES

Michéle Ménard and Kenneth M. Meyers

INTRODUCTION

Normal platelets contain four types of granules, including α granules, dense granules, lysosomes, and microperoxisomes. Platelet dense granules (also known as amine storage organelles, very dense granules, dense bodies, bull's-eye granules, and serotonin (5-HT) storage organelles) play a major role in primary hemostasis. Platelet secretion of dense granule constituents represents a positive feedback process to recruit more platelets at the site of vascular injury. Acquired and inherited platelet dense granule storage pool deficiencies (SPD) in humans and animal models are associated with bleeding tendencies. The dense granules accumulate and store nucleotides, divalent cations and amines in extremely high concentrations.[52] The bulk of the nucleotides stored in the dense granules are synthesized within megakaryocytes (MK), and only a small part is newly formed in circulating platelets.[12,25] Once platelets are into the circulation, they take up 5-HT produced by the enterochromaffin cells via two distinct transport systems: an imipramine-sensitive transporter in the plasma membrane and a reserpine-sensitive transporter in the dense granule membrane.[45] Within the granule, 5-HT is complexed with the nucleotides and divalent cations.[51] The formation of the storage complex provides osmotic stability to the dense granules and enhances the efficiency of 5-HT uptake.[3] This macromolecular complex within the dense granules can be demonstrated by transmission electron microscopy with glutaraldehyde fixation, to cross link the 5-HT within the complex, followed by osmium fixation.[49] When this type of fixation is used, the complex appears as a very electron-dense core of variable size.[49] Within bovine dense granules, the dense core is located eccentrically and appears to be attached to the luminal side of the granule membrane (Figure 1). The remaining content of the dense granule has a clear, empty-looking appearance.

Although the precursors of platelet dense granules are known to be synthesized within normal MK, their origin is still a matter of debate. It has been suggested that they derive from α granules,[53] the endoplasmic reticulum,[36] and the Golgi complex.[11] Within the MK, the dense granule precursors contain nucleotide/cation aggregates,[24,26] but lack 5-HT in amounts sufficient to be osmiophilic and acquire the characteristic appearance of platelet dense granules after glutaraldehyde fixation.[12,40,50] Osmiophilic dense granules are observed in

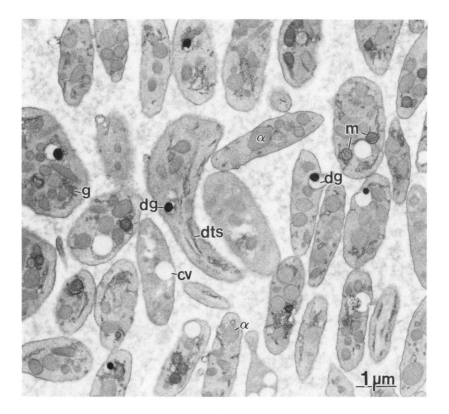

FIGURE 1. Bovine platelets fixed with glutaraldehyde to cross link the serotonin (5-HT), stored as macromolecular complexes with divalent cations and nucleotides within the dense granules, followed by osmium tetroxide postfixation. The macromolecular complexes appear as osmiophilic electron-dense cores of variable sizes located eccentrically within the dense granules (dg) where they appear to be attached to the luminal side of the granule membrane. The remaining contents of the dense granules take on a clear, empty-looking appearance. The clear vesicles (cv) could represent dense granule precursors without sufficient amounts of stored 5-HT to contain an osmiophilic dense core, or they may be sections of dense granules with a dense core situated at a different plane from the section. m, mitochondria; α, alpha granule; dts, dense tubular system; g, glycogen (magnification: × 11,200).

MK after *in vivo* and *in vitro* administration of 5-HT,[15,47,50] incubation with ethidium bromide,[25] or fixation in White's saline.[18,54] Dense granules also are revealed by electron microscopy of unfixed and unstained MK whole mount preparations.[24] Dense granule precursors can be demonstrated in MK at all stages of maturation using the uranaffin cytochemical reaction in which the uranyl ions react with the phosphate groups of the 5'-phosphonucleotides stored in these precursors.[11,44] The present chapter will discuss the ultrastructure of dense granule precursor development in bovine, canine, feline, and murine MK using standard fixation protocols and the following histochemical techniques: the uranaffin cytochemical reaction, calcium supplemented fixation, and studies with extracellular electron dense markers, including tannic acid, horseradish peroxidase, lanthanum and ruthenium red.

MK CLASSIFICATION WITH TRANSMISSION ELECTRON MICROSCOPY

MK were classified as immature, maturing, and mature according to standard criteria.[37] Immature MK are small cells (15 to 18 micron diameters) with a high nuclear:cytoplasmic ratio and few cytoplasmic organelles other than large mitochondria and numerous polyribosomes. Maturing MK are heterogeneous in size, have a lobulated nucleus, a well-developed Golgi complex, abundant rough endoplasmic reticulum, and variable numbers of α granules and demarcation membranes. Mature MK are heterogeneous in size, have a fully granulated cytoplasm with a fully developed demarcation membrane system (DM) and only remnants of the rough endoplasmic reticulum and Golgi complex.

DENSE GRANULE PRECURSORS IN MK WITH STANDARD PROCESSING FOR TRANSMISSION ELECTRON MICROSCOPY

Our standard protocol for transmission electron microscopy consists of a fixation in a solution of 2.5% glutaraldehyde in 0.1 M sodium cacodylate buffer pH 7.4, washes in 0.1 M sodium cacodylate buffer with 4.0% sucrose at pH 7.4, a post-fixation in a solution of 1.0% osmium tetroxide in the same buffer with 4.0% sucrose at pH 7.4, and an optional *en bloc* staining with a 1.0% uranyl acetate solution during dehydration in the 50% ethanol solution. When this protocol is used, dense granule precursors in MK appear as clear vesicles bounded by a sharp membrane[27] (Figures 2A and 3). A small membrane projection is commonly seen attached to the luminal side of the vesicle membrane and appears to be the site where the dense core is found when the uranaffin reaction is used.[27] These clear vesicles are easily differentiated from α granules by their clear, empty-appearing contents. They are differentiated from cross sections of the DM channels by their larger diameters, sharper membranes, and the absence of fine fibrillar material coating the internal side of their membranes. In addition, clear vesicles are differentiated from dilated channels in the rough endoplasmic reticulum by their clear contents, sharp membranes, and the absence of ribosomes attached to their membranes.[27]

The diameters of these clear vesicles are in the 250 to 650 nm range with a mean diameter ($\pm$ SD) of 400 nm $\pm$ 100 nm in cattle MK, 150 to 220 nm with a mean diameter ($\pm$ SD) of 180 nm $\pm$ 20 nm in mice MK, 200 to 440 nm with a mean diameter ($\pm$ SD) of 300 nm $\pm$ 60 nm in dog MK, 200 to 440 nm with a mean diameter ($\pm$ SD) of 340 nm $\pm$ 70 nm in cat MK. Daimon and David[11] have reported a range of 150 to 300 nm diameters for the clear vesicles observed

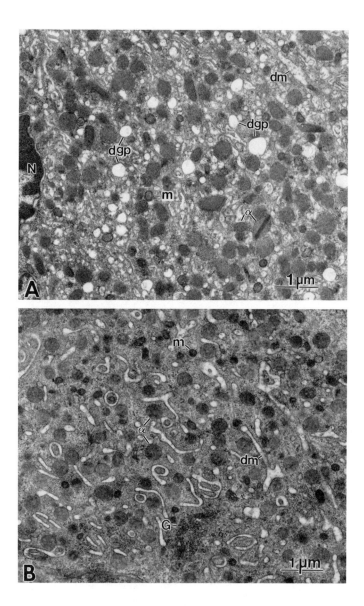

FIGURE 2. Mature normal bovine megakaryocytes (MK) processed for standard transmission electron microscopy. (A) Precursors of dense granules appear as clear, empty-looking vesicles (dgp), bounded by a sharp membrane, and are evenly distributed between the alpha granules (α). The demarcation membrane system (dm), α granules, clear vesicles and small mitochondria (m) comprise most of the cytoplasmic organelles. (B) Late maturing MK from CHS calf with standard processing. There is a virtual absence of clear vesicle precursors of dense granules. The demarcation membrane system, mitochondria and α granules are similar to those from normal MK. N, nucleus; G, Golgi complex (magnifications: A × 14,500; B × 14,500).

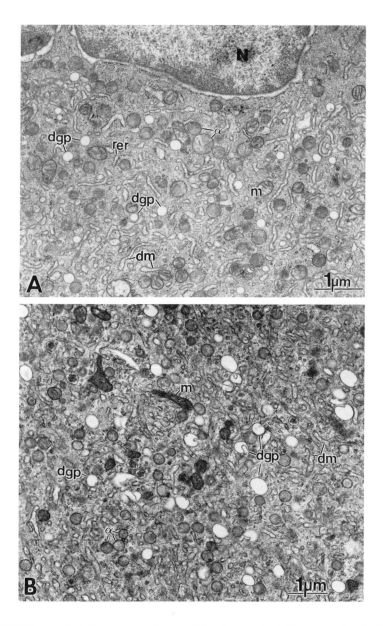

FIGURE 3. Maturing MK from a normal mouse (A), and a normal cat (B), processed for standard transmission electron microscopy. Precursors of dense granules appear as clear, empty-looking vesicles (dgp), bounded by a sharp membrane, and are evenly distributed between the alpha granules (α). Most of the dense granule precursors are smaller than the α granules in the mouse MK, whereas the opposite is the situation in the cat MK. A fair amount of rough endoplasmic reticulum (rer) is present in both the mouse and cat MK. N, nucleus; m, mitochondria; dm, demarcation membrane system (magnifications: A × 16,200; B × 13,800).

in rat MK. The dense granule precursors are larger in bovine MK than they are in other species, which is probably a reflection of the larger size of bovine platelet granules compared to platelet granules from other species.[31,58] It was surprising to find that the diameters of the dense granule precursors in dog and cat MK approached the diameters of dense granule precursors in bovine MK and were almost twice the size of the dense granule precursors in mice MK. This was especially surprising since dog and cat α granules were similar in size to those from mice MK and approximately half the size of bovine α granules. The size of the dense granule precursors in human MK has yet to be determined.

Precursors of the dense granules and α granules from different animal species appear to be synthesized simultaneously. Both types of granules are present early in the maturation process of normal MK and increase in number during maturation. In maturing and mature bovine MK, there is a wide range in the dense granule to α granule ratio (Figure 4A) with a mean of one clear vesicle per 8.8 (SD 3.58) α granules and a range of one clear vesicle per 3.75 to 16.06 α granules.[27] This wide variation in the concentration of dense granule precursors in normal bovine MK at similar levels of maturation supports the hypothesis that dense granule formation is a regulated process, as has been suggested for α granule proteins.[14] However, the control mechanisms for dense granule precursor synthesis by MK are unknown at this time.

Megakaryocytopoiesis involves humoral factors that regulate the proliferation of committed megakaryocytic progenitor cells and factors that regulate MK differentiation. Maturation of MK from immature to morphologically identifiable MK involves a number of processes, including the synthesis of cytoplasmic organelles, the acquisition of membrane antigens and glycoproteins, and the release of platelets. Recombinant cytokines, including interleukin (IL)-3, granulocyte-macrophage colony-stimulating factor, erythropoietin, IL-6, and IL-11, singly and/or in combination, have been shown to promote not only MK proliferation but also MK maturation *in vitro*, as defined by increased MK size, ploidy, maturational stage, enhanced expression of MK antigens and acetylcholinesterase.[7,8,19,21,22,23] IL-6 and, to a lesser extent IL-3, have also been shown to be a potent thrombopoietic factor *in vivo* in mice.[9,16,20] The effect of these cytokines on dense granule synthesis by MK is unknown at this time. Future experiments to elucidate these control mechanisms will be important to increase our understanding of dense granule precursor synthesis in normal and pathological conditions.

URANAFFIN CYTOCHEMICAL REACTION

The uranaffin cytochemical reaction of Richards and Da Prada[44] has been shown to be specific for the adenine nucleotide-storing organelles within platelets and MK. The formation of electron-dense cores within these granules is thought to be the result of an interaction between the uranyl ions and the phosphate groups of the nucleotides stored within the granules.[11,44] The mechanism by

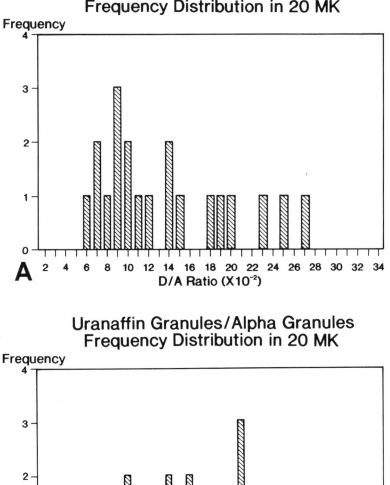

FIGURE 4. Dense granule precursor to α granule ratio frequency distribution in 20 maturing and mature MK from normal cattle (A). Uranaffin granule to α granule ratio frequency distribution in 20 maturing and mature MK from normal cattle (B)

which the membranes of the platelet dense granules and the dense granule precursors within the MK stain is poorly understood.[38] Uranyl ions may be reacting with phosphoproteins, phospholipids, nucleotides in transit across the membrane, other polyphosphate complexes, or a combination thereof.[38] To increase the sensitivity for the detection of the dense granule precursors within the MK, we prolonged the incubation time in the 4% aqueous uranyl solution to 30 hr.[27] When this technique is used in bovine MK, virtually all of the clear vesicles (dense granule precursors) are replaced by positive uranaffin granules of 100 to 700 nm diameters with strong staining of the membrane and/or of a dense core attached to the membrane (Figure 5B).

As observed with the clear vesicles after standard fixation, there is a wide range in the uranaffin granule to α granule ratios in both maturing and mature bovine MK (Figure 4B), with a mean of one uranaffin granule per 6.53 (SD 2.60) α granules and a range of one uranaffin granule per 3.06 to 14.52 α granules. The size discrepancy between the clear vesicles (250 to 650 nm diameters) and the uranaffin-positive granules (100 to 700 nm diameters), and the smaller number of clear vesicles as opposed to uranaffin granules, is due to the presence of small uranaffin-positive granules. These small vesicles are not counted as dense granule precursors in MK with standard fixation since, because of their small diameter, they cannot be differentiated from cross sections of the demarcation membrane system. Small uranaffin granules could represent peripheral cross sections of larger uranaffin granules or smaller granules that will eventually fuse and form the larger uranaffin granules. Serial sectioning has not yet been done to test either possibility.

Uranaffin granules are synthesized early in the maturation of bovine and feline MK and appear to be synthesized simultaneously with α granules. The most immature bovine and feline MK that we could identify all contained small numbers of both uranaffin-positive granules and α granules (Figure 5A). Whether these dense granule precursors are mature and able to store 5-HT is unknown. MK at all stages of maturity have been shown to accumulate 5-HT,[46] although electron microscopy was not done in those studies to determine if the 5-HT was stored in dense granule precursors or free in the cytoplasm. Electron microscopic studies using glutaraldehyde in White's saline and the uranaffin reaction have suggested that dense granule precursor maturation in human MK may occur at very late stages of MK maturation.[18,54]

FIXATION IN A CALCIUM-SUPPLEMENTED FIXATIVE

Adding up to 5.0 mM of calcium chloride to a fixative solution containing 2.5% glutaraldehyde in 0.1 M cacodylate buffer at pH 7.4 reveals dense granule precursors as clear vesicles similar to those observed with standard fixation for bovine, canine, and feline MK (Figures 6A, 6C, 6D). Using this type of fixation, there is a complete absence of osmiophilic dense granules in all the MK observed in those species. This contrasts with data from human MK in which

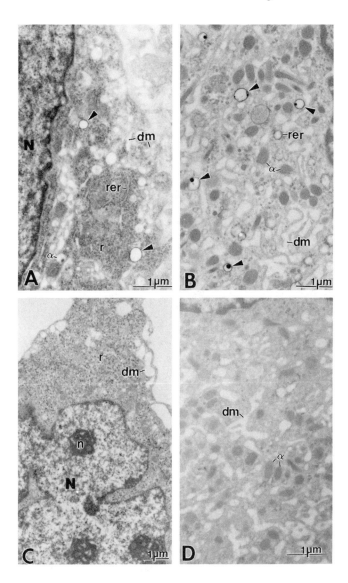

FIGURE 5. MK processed for the uranaffin cytochemical reaction. Immature MK from a normal cat (A) with a large number of polyribosomes (r) and rough endoplasmic reticulum (rer) and few alpha granules (α). Uranaffin-positive granules (arrow head) can already be seen at this early stage of development. Mature MK from a normal calf (B). The α granules and demarcation membranes (dm) are well developed. Numerous uranaffin granules (arrow heads) are present, with staining of the granule membrane and/or of a dense core attached to the membrane on the luminal side, and replace most of the clear vesicles observed using standard fixation. Immature (C), and mature (D) MK from a CHS calf. There is a virtual absence of uranaffin-positive granules or clear vesicles. The demarcation membrane system, α granules, and rough endoplasmic reticulum are similar to those from the normal MK. N, nucleus, n, nucleolus (magnification: A × 22,400; B × 13,500; C × 9,100; D × 14,000).

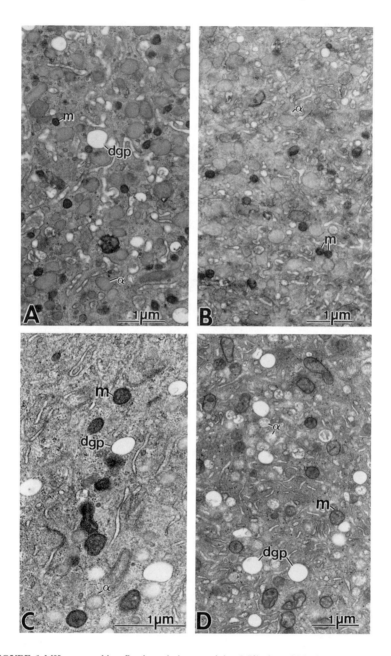

FIGURE 6. MK processed in a fixative solution containing 2.5% glutaraldehyde, 5.0 mM calcium chloride in 0.1 M sodium cacodylate buffer, pH 7.4. The dense granule precursors within MK from normal cattle (A), normal cats (C), and normal dogs (D) appear as clear, empty-looking vesicles (dgp) with a complete absence of osmiophilic dense cores. Such clear vesicles are completely absent in MK from CHS cattle (B). α, alpha granules; m, mitochondria (magnifications: A × 15,500; B × 14,000; C × 21,600; D × 15,600).

several osmiophilic dense granules are seen in mature MK fixed in a solution of 3.0% glutaraldehyde in White's saline.[18,54] The presence of osmiophilic dense granules has been attributed to the high calcium content of White's saline (3.2 mM). Human platelets contain large amounts of releasable calcium compared to other animal species[28] which may explain why osmiophilic dense granules are observed in this species after fixation in calcium-enriched fixative (White's saline solution). Alternatively, other components in White's saline solution may be responsible for the large number of osmiophilic dense granules in human MK. This would explain their absence in MK from cattle, dogs, and cats processed using 2.5% glutaraldehyde supplemented with 5 mM calcium in cacodylate buffer.

STUDIES WITH EXTRACELLULAR ELECTRON DENSE MARKERS

Tannic Acid

To delineate the extracellular environment and eliminate the possibility of any communication between the demarcation membrane system and the dense granule precursors, we added tannic acid to the fixative (2.0% glutaraldehyde, 2.0% tannic acid, 0.1 M cacodylate buffer, pH 6.8), washing buffer (1.0% tannic acid in 0.1 M cacodylate buffer, pH 7.0), and post-fixation solution (1.0% osmium tetroxide, 1.0% tannic acid, pH 7.4). We found no penetration of tannic acid into the dense granule precursors. Because tannic acid acted as a mordant for cytoplasmic constituents in bovine MK, giving them a very electron-dense appearance, the clear content of the dense granule precursors is even more evident than with standard fixation (Figure 7A).

Horseradish Peroxidase

Using horseradish peroxidase, we followed the technique described by Breton-Gorius[5] to delineate the DM of unfixed MK. The specimens were incubated in a balanced salt solution containing horseradish peroxidase at 4°C to inhibit membrane flow and endocytosis. We found that penetration of the DM was variable, with good delineation in some MK and a complete absence in others. In the MK with good delineation of the DM canaliculi, we found no horseradish peroxidase penetration into the dense granule precursors (Figure 7C).

Lanthanum and Ruthenium Red

Addition of lanthanum nitrate[48] or ruthenium red[57] to the fixatives and washing buffers to delineate the extracellular environment resulted in poor penetration of the DM canaliculi of bovine MK. Surprisingly, we found that the dense granule precursors contained dense cores and/or dark staining of the membrane similar to that observed with the uranaffin reaction in normal bovine MK (Figures 8A, 8C). In a process similar to the uranaffin reaction, small amounts of lanthanum, a trivalent cation, and ruthenium red, a hexavalent

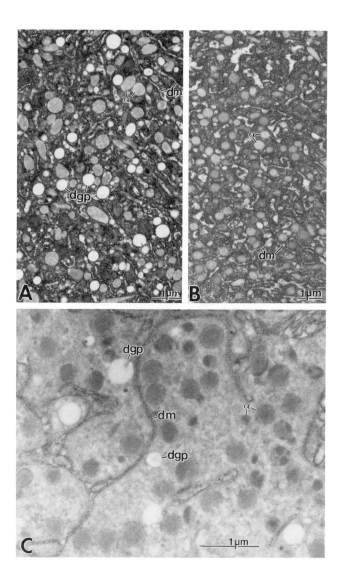

FIGURE 7. MK from a normal calf (A), and a CHS calf (B), processed with tannic acid, as an extracellular electron-dense marker, to delineate the demarcation membrane system (dm). There is a complete absence of tannic acid penetration into the dense granule precursors (dgp) of normal cattle MK (A). Tannic acid acts as a mordant for cytoplasmic constituents in bovine MK, resulting in an increased electron density of the cytoplasm and a strong contrast in the clear content of the dense granule precursors which are even more evident than with standard fixation. There is a complete absence of clear vesicles within MK from CHS cattle processed in a similar manner (B). Mature MK from a normal calf incubated with horseradish peroxidase prior to fixation (C). The demarcation membrane system is delineated by the diaminobenzidine product. There is a complete absence of penetration within the clear vesicle dense granule precursors. α, alpha granules (magnifications: A × 10,800; B × 9,500; C × 24,000).

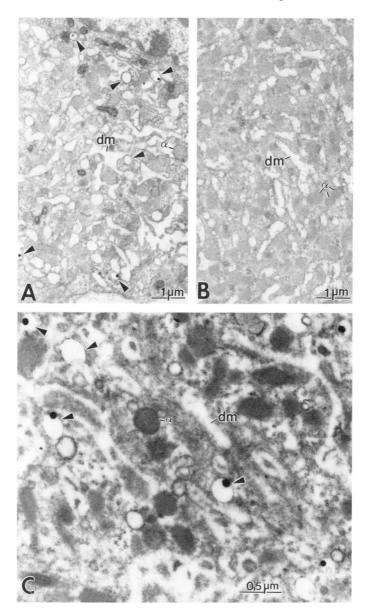

FIGURE 8. MK from a normal calf (A), and a CHS calf (B), processed with lanthanum nitrate in the fixative to delineate the demarcation membrane system (dm). There is poor penetration of the dm canaliculi. The dense granule precursors (arrow heads) contain dense cores and/or dark staining of the membrane similar to that observed with the uranaffin reaction for normal bovine MK. Such precipitates are completely absent in MK from CHS cattle. MK from a normal calf processed with ruthenium red in the fixative to delineate the dm (C). There is poor penetration of the dm canaliculi. The dense granule precursors (arrow heads) contain dense cores and/or dark staining of the membrane similar to what is observed with the uranaffin reaction in normal bovine MK. α, alpha granule (magnifications: A × 11,200; B × 11,200; C × 24,000).

cation may form electron-dense precipitates by their reaction with the phosphate groups of the nucleotides stored within the dense granule precursors. X-ray probe microanalyses must be conducted to confirm whether this is the case.

DENSE GRANULE PRECURSORS IN MK FROM ANIMAL MODELS OF THE CHEDIAK-HIGASHI SYNDROME (CHS)

The CHS is an autosomal recessive disorder characterized by enlarged cytoplasmic granules in most granule-forming cells, incomplete oculocutaneous albinism, increased susceptibility to pyogenic infections and a bleeding tendency.[41] This bleeding tendency is associated with a dense granule SPD in which there is a virtual absence of secretable ATP, ADP and 5-HT, and a substantial reduction of releasable cations.[1,2,4,6,10,29,30,34,35] The platelet SPD represents an unusual expression of the CHS trait since platelets from most CHS patients are virtually devoid of osmiophilic dense granules,[10,13,17,31,35,42,43] and enlarged granules are rarely observed.[13,31,35,55,56]

CHS Cattle

When MK from cattle with CHS are processed with standard fixation or with a calcium-supplemented fixative, there is a virtual absence of clear vesicles (dense granule precursors) in MK at all stages of maturity (Figures 2B, 6B).[27] With the uranaffin cytochemical reaction, there is a complete absence of uranaffin granules in all MK (Figures 5C, 5D).[27] Studies with extracellular electron-dense markers are also associated with a complete absence of clear vesicles (Figure 7B). When lanthanum nitrate is added to the fixatives, there is a complete absence of granules with electron-dense precipitates (Figure 8B). All the other organelles of CHS MK appear similar to those from normal bovine MK, whether standard fixation or cytochemical techniques are utilized.

The absence of platelet dense granule-related amine uptake processes[33] coupled with the virtual absence of stored adenine nucleotides,[29] an absence of dense granule specific membrane proteins,[32] and the absence of recognizable dense granule precursors within MK,[27] suggest that the platelet SPD associated with bovine CHS results from an anatomic deficiency of dense granules rather than a functional alteration whereby the granule is present but cannot perform processes required to obtain an amine/nucleotide/divalent cation storage complex. The absence of dense granule membrane proteins in membranes of other granules from bovine CHS platelets also suggests that fusion of dense granules or their precursors with other granules cannot account for the platelet dense granule deficiency in CHS platelets. The reason for this lack of synthesis of dense granule precursors within bovine CHS MK is not clear at this time.

CHS Cats

Preliminary studies with standard processing for transmission electron microscopy have shown that precursors of dense granules are virtually absent in feline CHS MK (Figure 9). Uranaffin granules are also virtually absent in MK

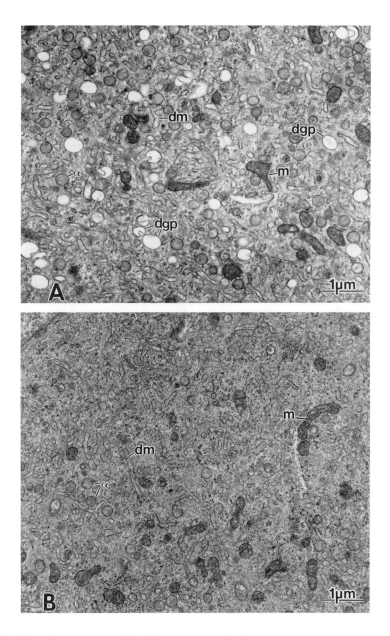

FIGURE 9. Normal (A) and CHS (B) feline maturing MK from bone marrow specimens processed for standard transmission electron microscopy. A large number of dense granule precursors (dgp) are present in the MK from the normal cat (A) and appear as clear vesicles bounded by sharp membranes. The demarcation membrane system (dm), rough endoplasmic reticulum, alpha granules (α) and mitochondria (m) comprise most of the other cytoplasmic organelles. Dense granule precursors are virtually absent in the feline CHS MK (B). The dm, m and α granules appear similar to those in normal feline MK. Enlarged granules are not seen (magnifications: A × 13,250; B × 13,250).

from CHS cats. No organelle abnormality other than the absence of dense granule precursors and uranaffin granules is observed in the MK from these animals. Numerous dense granule precursors and uranaffin granules are observed in all the MK sections from normal cat bone marrow specimens processed simultaneously. These findings indicate that the platelet dense granule SPD in CHS cats results from an anatomical absence of MK dense granule precursors rather than a functional alteration whereby the dense granule is present but cannot perform processes required to obtain an amine/nucleotide/divalent storage complex. These results agree with our findings in CHS cattle. The absence of platelet dense granules in feline CHS also appears to result from a lack of synthesis of dense granule precursors by the MK, but the reason feline CHS MK do not synthesize dense granule precursors is not clear at this time.

Beige (C57BL/6J-bg^J/bg^J) Mice

Preliminary data with MK from beige mice indicate some important differences between this animal model of CHS and CHS cats and cattle (Figure 10). In contradistinction with feline and bovine CHS MK which virtually lack precursors of dense granules, MK sections from beige mice have a marked reduction of precursors of dense granules, although they contain a few normal-appearing dense granule precursors when processed for standard transmission electron microscopy and the uranaffin cytochemical reaction. Uranaffin granules were not identified in previous reports using the uranaffin cytochemical reaction and beige mice MK, although these investigators used a shorter incubation period with uranium ions.[44] As opposed to feline and bovine CHS MK in which enlarged granules were not observed, enlarged granules with contents of similar electron density to that of α granules or lysosomes are common in beige mice MK processed for standard transmission electron microscopy. Preliminary results indicate that some of these enlarged granules are positive for acid phosphatase cytochemistry and are, therefore, enlarged lysosomes. These findings show that heterogeneity is present among the animal models of CHS as is observed with human patients.[39] They also suggest that SPD probably result from different defects in the synthesis of dense granule precursors. These different animal models should be very useful to dissect dense granule precursor synthesis within MK and to better understand the pathobiology of these granules and the pathogenesis of storage pool deficiencies.

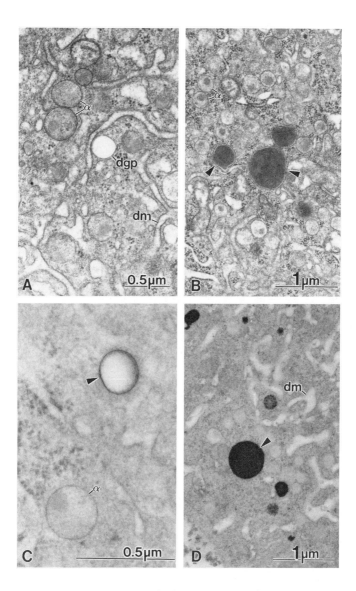

FIGURE 10. MK from beige (C57BL/6J-bg^J/bg^J) mice processed for standard transmission electron microscopy (A) and (B), the uranaffin cytochemical reaction (C), and acid phosphatase cytochemistry (D). Although there is a marked reduction of precursors of dense granules, a few normal-appearing dense granule precursors (dgp) are present in MK processed for standard transmission electron microscopy (A), and these stain positively with the uranaffin cytochemical reaction (arrow head) (C). Enlarged granules with contents of similar electron density to that of α granules or lysosomes are common in beige mice MK processed for standard transmission electron microscopy (arrow heads) (B). Some of these enlarged granules are positive for acid phosphatase cytochemistry and are, therefore, enlarged lysosomes (arrow head) (D). dm, demarcation membrane system; α, alpha granule (magnifications: A × 33,600; B × 22,400; C × 70,000; D × 22,400).

List of Abbreviations
5-HT = Serotonin
CHS = Chediak-Higashi syndrome
DM = Demarcation membrane system
IL = Interleukin
MK = Megakaryocytes
SPD = Storage pool deficiency

REFERENCES

1. **Apitz-Castro, R., Cruz, M.R., Ledezma, E., Merino, F., Ramirez-Duque, P., Dangelmeier, C., and Holmsen, H.**, The storage pool deficiency in platelets from humans with the Chediak-Higashi syndrome: Study of six patients, *Br. J. Haematol.*, 59, 471-483, 1985.
2. **Bell, T.G., Meyers, K.M., Prieur, D.J., Fauci, A.S., Wolff, S.M. and Padgett, G.A.**, Decreased nucleotide and serotonin storage associated with defective function in Chediak-Higashi syndrome cattle and human platelets, *Blood*, 48, 175-184, 1976.
3. **Berneis, K.H., Da Prada, M., and Pletsher, A.**, Micelle formation between 5-hydroxytamine and adenosine triphosphate in platelet storage organelles, *Nature*, 165, 913-914, 1969.
4. **Boxer, G.J., Holmsen, H., Robkin, L., Bang, N.U., Boxer, L.A., and Baehner, R.L.**, Abnormal platelet function in the Chediak-Higashi syndrome, *Br. J. Haematol.*, 35, 521-533, 1977.
5. **Breton-Gorius, J.**, Development of two distinct membrane systems associated in giant complexes in pathological megakaryocytes, *Ser. Haemat.*, 8, 49-67, 1975.
6. **Buchanan, G.R., and Handin, R.I.**, Platelet function in the Chediak-Higashi syndrome, *Blood*, 47, 941-948, 1976.
7. **Burstein, S.A.**, Interleukin 3 promotes maturation of murine megakaryocytes *in vitro*, *Blood Cells*, 11, 469-474, 1986.
8. **Burstein, S.A., Mei, R., Henthorn, J., Friese, P., and Turner, K.**, Recombinant human leukemia inhibitory factor (LIF) and interleukin-11 (IL-11) promote murine and human megakaryocytopoiesis *in vitro*, *Blood*, 76, 450a, 1990.
9. **Carrington, P.A., Hill, R.J., Stenberg, P.E., Levin, J., Corash, L., Schreurs, J., Baker, G., and Levin, F.C.**, Multiple *in vivo* effects of interleukin-3 and interleukin-6 on murine megakaryocytopoiesis, *Blood*, 77, 34-41, 1991.
10. **Costa, J.L., Fauci, A.S., and Wolff, S.M.**, A platelet abnormality in the Chediak-Higashi syndrome of man, *Blood*, 48, 517-520, 1976.
11. **Daimon, T., and David, H.**, Precursors of monoamine-storage organelles in developing megakaryocytes of the rat, *Histochemistry*, 77, 353-363, 1983.
12. **Da Prada, M., Richards, J.G., and Kettler, R.**, Amine storage organelles in platelets, in *Platelets in Biology and Pathology*, Vol. 5, Gordon, A.S., Ed., Elsevier/North-Holland Biomedical Press, Amsterdam, 1981, 107-145.
13. **Fagerland, J.A., Hagemoser, W.A., and Ireland, W.P.**, Ultrastructure and sterology of leukocytes and platelets of normal foxes and a fox with a Chediak-Higashi-like syndrome, *Vet. Pathol.*, 24, 164-169, 1987.
14. **Gewirtz, A.M., Keefer, M., Doshi, K., Annamalai, A.E., Chong Chiu, H., and Colman, R.W.**, Biology of human megakaryocyte factor V, *Blood*, 67, 1639-1648, 1987.
15. **Hagen-Aukamp, C., Wasemann, W., and Aumuller, G.**, Intracellular distribution of adenine and 5-hydroxytryptamine in megakaryocytes isolated by density gradient and velocity sedimentation from bone marrow, *Eur. J. Cell Biol.*, 23, 149-156, 1980.
16. **Hill, R.J., Warren, M.K., Stenberg, P., Levin, J., Corash, L., Drummond, R., Baker, G., Levin, F., and Mok, T.**, Stimulation of megakaryocytopoiesis in mice by human recombinant interleukin-6, *Blood*, 77, 42-48, 1991.

17. **Holland, J.M.**, Serotonin deficiency and prolonged bleeding in beige mice, *Proc. Soc. Exp. Biol. Med.*, 151, 32-39, 1976.

18. **Hourdille, P., Fialon, P., Belloc, F., Boisseau, M.R., and Andrieu, J.M.**, Mepacrine labeling test and uranaffin cytochemical reaction in human megakaryocytes, *Thromb. Haemost.*, 47, 232-235, 1982.

19. **Ishibashi, T., and Burstein, S.A.**, Interleukin 3 promotes the differentiation of isolated single megakaryocytes, *Blood*, 67, 1512-1514, 1986.

20. **Ishibashi, T., Kimura, H., Shikama, Y., Uchida, T., Kariyone, S., Hirano, T., Kishimoto, T., Takatsuki, F., and Akiyama, Y.**, Interleukin-6 is a potent thrombopoietic factor *in vivo* in mice, *Blood*, 74, 1241-1244, 1989.

21. **Ishibashi, T., Kimura, H., Uchida, T., Kariyone, S., Friese, P., and Burstein, S.A.**, Human interleukin 6 is a direct promotor of maturation of megakaryocytes *in vitro*, *Proc. Natl., Acad. Sci. USA*, 86, 5953-5957, 1989.

22. **Ishibashi, T., Koziol, J.A., and Burstein, S.A.**, Human recombinant erythropoietin promotes differentiation of murine megakaryocytes *in vitro*, *J. Clin. Invest.*, 79, 286-289, 1987.

23. **Ishibashi, T., Ruggeri, Z.M., Harker, L.A., and Burstein, S.A.**, Separation of human megakaryocytes by state of differentiation on continuous gradients of Percoll: Size and ploidy of cells identified by monoclonal antibody to glycoprotein IIb/IIIa, *Blood*, 67, 1286-1292, 1986.

24. **Levine, R.F.**, Megakaryocyte biochemistry, in *Biochemistry of Platelets*, Academic Press, San Diego, 1986, 417-442.

25. **Levine, R.F., and Costa, J.L.**, Interaction of ethidium bromide with serotonin storage granules in platelets and megakaryocytes, *J. Cell Biol.*, 79, 376a, 1978.

26. **Levine, R.F., and Webster, H.K.**, Purine metabolism in megakaryocytes and platelets, *Thromb. Haemost.*, 46, 227, 1981.

27. **Ménard, M., and Meyers, K.M.**, Storage pool deficiency in cattle with the Chediak-Higashi syndrome results from an absence of dense granule precursors in their megakaryocytes, *Blood*, 72, 1726-1734, 1988.

28. **Meyers, K.M., Holmsen, H., and Seachord, C.L.**, Comparative study of platelet dense granule constituents, *Am. J. Physiol.*, 243, R454-R461, 1982a.

29. **Meyers, K.M., Holmsen, H., Seachord, C.L., Hopkins, G.E., Borchard, R., and Padgett, G.A.**, Storage pool deficiency in platelets from Chediak-Higashi cattle, *Am. J. Physiol.*, 237, R239-R248, 1979.

30. **Meyers, K.M., Holmsen, H., Seachord, G., Hopkins, C.L., and Gorham, J.**, Characterization of platelets from normal mink and mink with the Chediak-Higashi syndrome, *Am. J. Hematol.*, 7, 137-146, 1979.

31. **Meyers, K.M., Hopkins, G., Holmsen, H., Benson, K., and Prieur, D.J.**, Ultrastructure of resting and activated storage pool deficient platelets from animals with the Chediak-Higashi syndrome, *Am. J. Pathol.*, 106, 364-377, 1982.

32. **Meyers, K.M., and Seachord, C.**, Identification of dense granule specific membrane proteins in bovine platelets that are absent in the Chediak-Higashi syndrome, *Thromb. Haemost.*, 64, 319-325, 1990.

33. **Meyers, K.M., Seachord, C.L., Benson, K., Fukami, M., and Holmsen, H.**, Serotonin accumulation in granules of storage pool-deficient platelets of Chediak-Higashi cattle, *Am. J. Physiol.*, 245, H150-H158, 1983.

34. **Meyers, K.M., Seachord, C.L., Holmsen, H., and Prieur, D.J.**, Evaluation of the platelet storage pool deficiency in the feline counterpart of the Chediak-Higashi syndrome, *Am. J. Hematol.*, 11, 241-253, 1981.

35. **Parmley, R.T., Poon, M.C., Crist, W.M., and Malluh, A.**, Giant platelet granules in a child with the Chediak Higashi syndrome, *Am. J. Hematol.*, 6, 51-60, 1979.

36. **Parmley, R.T., Spicer, S.S., and Wright, N.J.**, Diaminobenzidine activity in the platelet dense body, *J. Histochem. Cytochem.*, 22, 1063-1067, 1974.

37. **Paulus, J.M.**, DNA metabolism and development of organelles in guinea-pig megakaryocytes: A combined ultrastructural, autoradiographic and cytophotometric study, *Blood,* 35, 298-311, 1970.

38. **Payne, C.M.**, A quantitative ultrastructural evaluation of the cell organelle specificity of the uranaffin reaction in normal human platelets, *Am. J. Clin. Pathol.,* 81, 62-70, 1984.

39. **Penner, J.D., and Prieur, D.J.**, Interspecific genetic complementation analysis with fibroblasts from humans and four species of animals with Chediak-Higashi syndrome, *Am. J. Med. Genet.,* 28, 455-470, 1987.

40. **Pletscher, A., and Da Prada, M.**, The organelles storing 5-hydroxytryptamine in blood platelets, in *Biochemistry and Pharmacology of Platelets,* Elsevier/Excerpta Medica/North-Holland, Amsterdam (Ciba Foundation Symposium 35), 1975, 261-279.

41. **Prieur, D.J., and Collier, L.L.**, Chediak-Higashi syndrome of animals, *Am. J. Pathol.,* 90, 533-536, 1978.

42. **Prieur, D.J., Holland, J.M., Bell, T.G., and Young, D.M.**, Ultrastructural and morphometric studies of platelets from cattle with the Chediak-Higashi syndrome, *Lab. Invest.,* 35, 197-204, 1976.

43. **Rendu, F., Breton-Gorius, J., Lebret, M., Klebanoff, C., Buriot, D., Griscelli, C., Levy-Toledano, S., and Caen, J.P.**, Evidence that abnormal platelet functions in human Chediak-Higashi syndrome are the result of a lack of dense bodies, *Am. J. Pathol.,* 111, 307-314, 1983.

44. **Richards, J.G., and Da Prada, M.**, Uranaffin reaction: A new cytochemical technique for the localization of adenine nucleotides in organelles storing biogenic amines, *J. Histochem. Cytochem.,* 25, 1322-1336, 1977.

45. **Rudnick, G., Fishkes, H., Nelson, P.J., and Schuldiner, S.**, Evidence for two distinct serotonin transport systems in platelets, *J. Biol. Chem.,* 255, 3638-3641, 1980.

46. **Schick, P.K., and Weinstein, M.**, A marker for megakaryocytes: Serotonin accumulation in guinea pig megakaryocytes, *J. Lab. Clin. Med.,* 98, 607-615, 1981.

47. **Seitz, R., and Wasemann, W.**, Studies on megakaryocytes: Isolation from rat and guinea pig and incorporation of 5-hydroxytryptamine, *Eur. J. Cell Biol.,* 21, 183-187, 1980.

48. **Shaklai, M., and Tavassoli, M.**, Demarcation membrane system in rat megakaryocyte and the mechanism of platelet formation: A membrane reorganization process, *J. Ultrastruct. Res.,* 62, 270-285, 1978.

49. **Tranzer, J.P., Da Prada, M., and Pletscher, A.**, Ultrastructural localization of 5-hydroxytryptamine in blood platelets, *Nature,* 212, 1574-1575, 1966.

50. **Tranzer, J.P., Da Prada, M., and Pletscher, A.**, Storage of 5-hydroxytryptamine in megakaryocytes, *J. Cell Biol.,* 52, 191-197, 1972.

51. **Ugurbil, K.**, Studies on storage mechanisms in the dense granules: Nuclear magnetic resonance, in *Platelet Responses and Metabolism, Receptors and Metabolism,* Vol. 11, Holmsen, H., Ed., CRC Press, Boca Raton, Florida, 1987, 153-170.

52. **Ugurbil, K., Fukami, M.H., and Holmsen, H.**, ^{31}P NMR studies of nucleotide storage in the dense granules of pig platelets, *Biochemistry,* 23, 409-416, 1984.

53. **White, J.G.**, The dense bodies of human platelets: Origin of serotonin storage particles from platelet granules, *Am. J. Pathol.,* 53, 791-808, 1968.

54. **White, J.G.**, Serotonin storage organelles in human megakaryocytes, *Am. J. Pathol.,* 63, 403-408, 1971.

55. **White, J.G.**, Platelet microtubules and giant granules in the Chediak-Higashi syndrome, *Am. J. Med. Technol.,* 44, 273-278, 1978.

56. **White, J.G., and Gerrard, J.M.**, Ultrastructural features of abnormal blood platelets, *Am. J. Pathol.,* 83, 590-614, 1976.

57. **Wight, T.N., and Ross, R.**, Proteoglycans in primate arteries. I. Ultrastructural localization and distribution in the intima, *J. Cell Biol.,* 67, 660-674, 1975.

58. **Zucker-Franklin, D., Benson, K.A., and Meyers, K.M.**, Absence of a surface connected canalicular system in bovine platelets, *Blood,* 65, 241-244, 1985.

Chapter 3

NUCLEAR MAGNETIC RESONANCE STUDIES OF AMINE AND NUCLEOTIDE STORAGE MECHANISMS IN PLATELET DENSE GRANULES

Holm Holmsen and Kamil Ugurbil

INTRODUCTION

A most characteristic feature of the anucleate platelet is the presence of a large number of secretory storage granules. Contents of these granules are secreted upon cell stimulation, making the platelet a typical secretory cell. Storage granules are of three types, distinguishable both by their different physical appearance and by their contents. The α-granules and lysosomes resemble storage granules in other secretory cells and contain proteins (coagulation factors, glycoproteins, platelet-specific proteins) and hydrolases (particularly glycosidases), respectively.[23, 30, 36] The dense granules are inherently electron dense, very osmiophilic, and have the highest specific density of all platelet organelles.[17] They contain divalent cations, nucleotides and amines in proportions that vary considerably between species.[24] The substances secreted from storage granules are important in hemostasis and thrombosis, making intact storage and secretory mechanisms important for normal platelet function. This importance has been clearly demonstrated in patients with an inherited deficiency of the dense granule contents.[21]

This chapter reviews the storage mechanism(s) of the dense granule contents. Most of our knowledge about these mechanisms stems from NMR studies of intact platelets and isolated platelet granules from pig. These studies offer a plausible explanation for the osmotic stability of dense granules, which contain nucleotides and divalent metals in the MOLAR range, creating apparent concentration gradients several thousandfold over their surrounding membranes.

NMR MEASUREMENTS: SOME BASIC CONSIDERATIONS

Magnetic resonances, measured by a NMR spectrometer as energy in the radiofrequency range, are due to the distinct nuclear spin for each nucleus type. Thus, nuclei commonly found in biomolecules with nuclear spin are those with odd mass numbers such as ^{1}H, ^{13}C, and ^{31}P; they have widely separated NMR frequencies. (Note that ^{16}O, ^{12}C and ^{32}S, which also are common in biomolecules, have no nuclear spin.) However, the NMR frequency for a given nucleus is dependent on the chemical environment of that nucleus, due to the electron shielding of the nucleus from the influence of the spectrometer's external magnetic field. This shielding effect depends greatly on the electron density around the nucleus and, therefore, on the type of chemical bonding that governs an atom in a molecule. Electronic shielding is usually measured as chemical shift, which means that NMR frequency corresponding to a given degree of shielding is specific for a specific chemical bond.

Chemical shifts due to different chemical bonds are shown in Figure 1 and demonstrate a ^{31}P-NMR spectrum of pure ADP, ATP and P_i in dilute aqueous solution at pH 7.4. The P in the terminal phosphoryl groups of ATP and ADP (i.e., the γ-phosphoryl in ATP and the β-phosphoryl in ADP) have an electron environment made up of the bond to an O bound to a P (acid anhydride), the double bond to an O and two single bonds to O of dissimilar degree of protonation due to pH. We see in Figure 1 that the chemical shift of these two P atoms with a nearly identical electronic environment is quite similar. Furthermore, the P in the α-phosphoryl groups of ATP and ADP has almost the same chemical shifts (Figure 1) but is different from that of the P in the terminal phosphoryls, since this P has a similar electron environment made up of one ester bond, one anhydride bond, one double bond to O and a single bond to O. The P in the β-phosphoryl group in ATP has a chemical shift quite different from the two kinds of electron-shielded Ps mentioned above (Figure 1), since this P has two anhydride bonds, one double bond to O, and one single bond to O. Naturally, P in P_i, with one double bond to O and three single bonds to Os of different protonation, has a chemical shift entirely different from the P atoms with the electronic environment described above (Figure 1).

In general, the area under a peak is proportional to the concentration of the nucleus in a given molecular arrangement and given microenvironment. However, especially for the nucleotides and serotonin in the platelet granules, this is not always apparent in the spectra because the interaction between molecules varies depending on temperature and other factors.

Due to pH in the microenvironment of the molecules, the P in P_i, and in the terminal phosphoryl groups in ATP and ADP, are bound to O of different degrees of protonation. The chemical shifts of these Ps are, therefore, dependent on pH, while that of P in the α- and β-phosphoryls of ATP are not (Figure 2). The chemical shift of the Ps in ATP and ADP is also affected by binding of divalent metal ions to the polyphosphate moieties of the nucleotides.

The chemical shifts discussed above are due to intramolecular electron shielding, yet to appreciate the results described below, one must also realize that intermolecular electron shielding affects the chemical shift. This is illustrated in Figure 3A which shows the effect on ^{31}P chemical shifts of ATP and ADP in the presence of Mg^{2+} as a function of the concentration of the nucleotides with a fixed nucleotide/metal ratio. As will be discussed below, these changes are due to aggregation of the (aromatic) adenine rings from different ATP and ADP molecules. A similar intermolecular effect on the chemical shift of the H-atoms in the indole ring of serotonin, when complexing with ATP, is shown in Figure 3B.

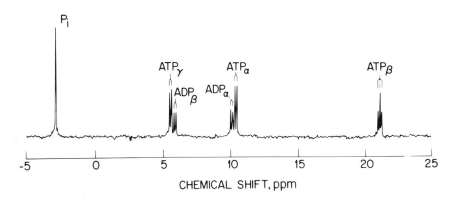

FIGURE 1. ^{31}P-NMR spectrum (145.7 MHz) of an aqueous solution of 4.5 mM ATP, 2.5 mM ADP and 1 mM P_i at pH 7.45. See text for explanation.[34]

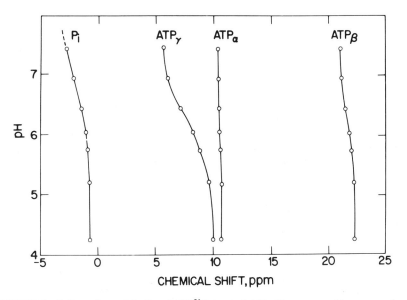

FIGURE 2. pH-Dependence of the P_i and ATP ^{31}P chemical shifts. The same conditions are used as in Figure 1 (4°C).[34]

COMPARTMENTATION OF NUCLEOTIDES IN PLATELETS

Platelets are rich in nucleotides, ADP and ATP in particular. Two-thirds of the latter are sequestered in the dense granules in a metabolically inert form, while the remainder is present in the cytoplasm and participates actively in metabolism.[16] These adenine nucleotide compartments will be referred to as the granule and the cytoplasmic adenine nucleotides, respectively. The cytoplasmic adenine

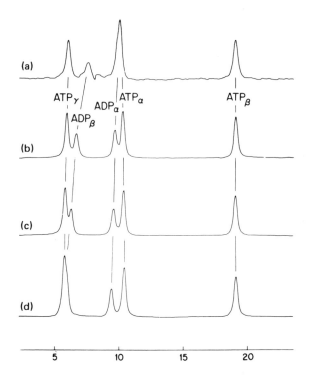

FIGURE 3A. Intermolecular electron shielding effects on chemical shift. Concentration of nucleotides. ^{31}P-NMR spectra of ATP, ADP and Mg^{2+} mixture as a function of solute concentration at pH 5.3 and 35°C. Both the ATP/ADP and the Mg^{2+}/(ATP + ADP) ratios were maintained at 2. The ATP concentrations were 4.5 mM (a); 45 mM (b); 90 mM (c); 270 mM (d). The line widths here are broader than in Figure 1 because the spectra were processed with a large (20 Hz) exponential filter.[34]

nucleotides are readily labeled with isotopic precursors,[16] and their concentrations are rapidly and specifically lowered either by metabolic inhibitors such as 2-deoxyglucose plus antimycin A[19] or by elevation of the hexose monophosphate shunt rate by H$_2$O$_2$ in the presence of azide.[18] Both treatments transfer the energy-rich phosphates in ATP and ADP to glycolytic intermediates.[20, 37]

Cytoplasmic ADP is further distributed between actin-bound and free ADP in a 2:3 proportion; actin-bound ADP is insoluble in 45% ethanol but extracted by perchloric acid.[11] The actin-bound ADP is freely exchangeable with the free ADP (i.e., it is labeled by isotopic precursors) but not readily available to ADP-consuming cellular processes.[12]

CONTENTS OF DENSE GRANULES IN HUMAN AND PORCINE PLATELETS

Table 1 shows the different constituents of human and porcine platelet-dense granules and the calculated concentrations within the granules. In both species the overall concentration of the solutes is 2-3 M, which implies that an osmotic

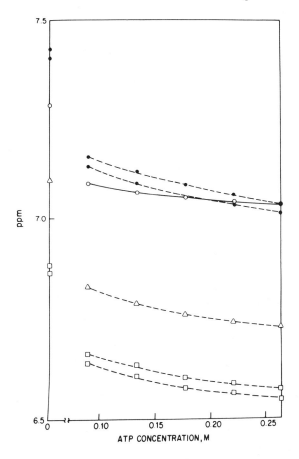

FIGURE 3B. Interaction between serotonin and ATP. Shown are the chemical shifts of the aromatic hydrogens of serotonin as a function of the ATP concentration at 35°C. H7 (●), H2 (o), H4 (Δ) and H6 (□) resonances from 5 mM at pH 5.1. The H7 and H6 appear as well-resolved doublets due to scalar coupling to each other; both peaks of the doublet are plotted for each resonance.[33] (Reprinted with permission from *Biochemistry.* 23, copyright 1984, American Chemical Society.)

gradient of 7-10 must exist across the granule membrane assuming that the osmolarity of the cytoplasmic environment is 300 mOsmol and the granule constituents are free in solution. Such a large osmotic imbalance will not be tolerated by any biological membrane, and as we will see from the NMR studies below, the constituents are present in aggregate complexes which effectively lower the interior osmolarity of the granules. It is also evident that ATP and ADP are the most abundant nucleotides in the granules from both species and that guanine nucleotides are one-tenth as abundant (Table 1).

Almost exclusively, porcine granules have Mg^{2+} and human granules have Ca^{2+} as the major divalent metal ions (Table 1). This difference is of vital importance for the appearance of the NMR signals from the granule contents, as will be discussed below. Another difference is the much higher content of ADP relative to ATP in human as opposed to porcine granules.

TABLE 1

Contents of Human and Porcine Dense Granules

Intragranular concentration $(mM^a)^b$

	Ca^{2+}	Mg^{2+}	ATP	ADP	GTP	GDP	PP_i	5-HT
Human	2181	<100	436	653	63	18	236	65
Porcine	<50	1370	443	255	46	13	44	180

[a]Calculations of concentrations are based on the assumption that there are 7 dense granules per platelet, all spherical with an average diameter of 198 nm.[9]
[b]The data are taken from Holmsen and Weiss[21] (human) and from the analysis in our laboratory (porcine).

USE OF ^{31}P-NMR TO STUDY NUCLEOTIDES IN INTACT PLATELETS AND ISOLATED GRANULES

Studies on Intact Human Platelets and Their Extracts

Figure 4 (a) and (b), respectively, show the phosphorus resonances of intact human platelets and their demineralized perchloric extract. The resonances from ATP and ADP in the extract are very similar to those obtained with the pure nucleotides, shown in Figure 1 with sharp, split bands, while those from the intact cells are very wide and have no splitting. This is due to different protonization and binding to Mg^{2+} for the single molecules, both intracellularly and intercellularly, as well as to the magnetic field heterogeneity in the NMR samples which were not spun. The different ATP-β resonances in spectra from cells and extracts reflect the binding of ATP to Mg^{2+} in the cells but not in the extract. Another difference between spectra from intact cells and from the extract is the clear resolution of the resonances of the terminal phosphates in ATP and ADP (γ- and β-phosphates, respectively) in the extract but not in the intact cell (Figure 4, 5.1 ppm).

The relative concentrations of ATP and ADP expressed as the ATP/(ATP + ADP) ratio can be determined from the spectra in Figure 4 as the ratio between the area under the ATP-β peak (19.0 ppm) and that of the combined terminal phosphates of ATP and ADP at 5.1 ppm. For several platelet suspensions the ATP/(ATP + ADP) was 0.95 + 0.1 in the intact cells but only 0.59 in the extract. The latter ratio accords well with that (0.63) obtained by chemical measurements of the extract. The (NMR) ratio obtained in intact cells, however, was slightly greater (0.83) than that found in perchloric extracts for the nucleotides that were labeled by ^{14}C-adenine in the same (intact) platelets used for the NMR experiment shown in Figure 4.[35] In ethanol extracts of such labeled platelets, lacking the F-actin-bound (labelable) ADP, the ATP/(ATP + ADP) ratio is 0.89 + 0.06, N = 7;[10] thus, it is very close to the NMR ratio from intact cells.

Ratio differences between the NMR spectra of intact platelets and their extract suggested: 1) that considerable amounts of ADP, demonstrable in the extracts by

NMR (i.e., after cell disruption), were not detectable by NMR in the intact human platelets; and 2) that only the cytoplasmic nucleotides, except F-actin-bound ADP, gave demonstrable NMR signals. It follows from these indications that the granule- and F-actin-bound ADP in human platelets do not give detectable NMR signals.

If these assumptions were correct, specific transfer of phosphate from the cytoplasmic adenine nucleotides to glycolytic intermediates by H_2O_2 in the presence of N_3^- (see above) should cause complete disappearance of the γ- and β-phosphate resonances in intact platelets. There would also be a corresponding appearance of monoesterphosphate resonances since ATP and ADP are converted to a mixture of IMP and AMP, i.e., monoesterphosphates,[18] and their γ- and β-phosphates are trapped as glycolytic intermediates, which are also monoesterphosphates. The spectra (a) and (b) in Figure 5 show that this is indeed the case. All ATP resonances in the intact platelets disappear, and a huge peak of

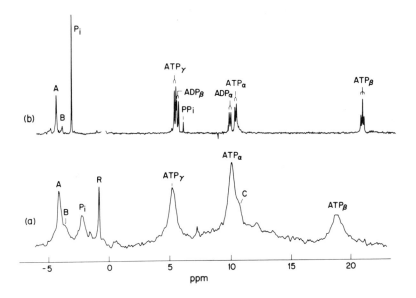

FIGURE 4. ^{31}P-NMR spectra (145.7 MHz) of intact human platelets (a) and their extract (b) at 4°C. Human platelets were washed at 4°C according to Haslam[15] and suspended at about 40 mg protein/ ml with 0.13 M NaCl/0.02 M TrisHCl/0.003 M EDTA/15 mM glucose. The spectrum (a) of this suspension is the sum of 4000 scans (40° pulses, 0.34 sec repetition time). The extract was obtained by addition of $HClO_4$ (final conc. 1 M) to the same cells shown in (a), followed by removal of debris by centrifugation, removal of divalent cations by passage through a Chelex-100 column, lyophilization and final resuspension with H_2O in the same volume as the original platelet suspension (pH adjusted to 7.5). The extract spectrum (b) is the sum of 2500 scans (40° pulses, 6 sec repetition times). A and B are unidentified resonances, probably from monoesterphosphates. C is probably resonances from granule ATP-α. R is a reference, 0.1% phosphoric acid in 0.1 N HCl contained in a concentric capillary.[35] (Originally published in *Proceedings* of National Academy of Sciences.)

monoesterphosphate appears after treatment with H_2O_2/N_3^-. Such treated platelets should contain adenine nucleotides only in the dense granules, and as seen in Figure 5, spectrum (b), they do not give NMR signals. Exactly the same results were obtained by us (data not shown) and by Costa *et al.*,[6] when the cytoplasmic adenine nucleotides were removed by incubation with 2-deoxyglucose + Antimycin A. To prove that the granule nucleotides were indeed present in the ATP + ADP-exhausted platelets, $HClO_4$ was added to these platelets. The NMR spectrum of the crude extract obtained (pH = O) was taken and showed distinct ATP/ADP resonances (Figure 5, spectrum (c)). After the debris and divalent

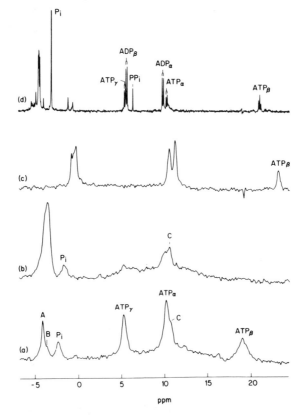

FIGURE 5. Effect of H_2O_2/N_3^- on human platelets; ^{31}P-NMR analyses of intact cells and extracts. Suspensions of human platelets were prepared as described in Figure 4. Shown are ^{31}P-NMR spectra (6000 scans) at 145.7 MHz at 4°C before (a) and after (b) treatment of the cells for 3 min at 25°C with 1.4 mM H_2O_2 and 1mM NaN_3. The spectrum in (c) was obtained (4000 scans, pH = 0) after addition of 60% $HClO_4$ to the sample in (b), and that in (d) was obtained (1800 scans) after the debris and divalent cations were removed by centrifugation and Chelex-100, respectively, from the extract in (c) and the pH was adjusted to 7.5. All spectra were measured at 40° pulses and repetition times of 0.34 sec (a,b and c) and 2.8 sec (d). The monoesterphosphate peak at -3.7 ppm in (b) is a composite resonance from IMP, AMP and glycolytic intermediates. The two resonances at 10.5 and 11.5 ppm in (c) stem from the α- and terminal phosphates of ATP and ADP and from PP_i.[35] (Originally published in *Proceedings* of National Academy of Sciences.)

cations were removed from this extract and its pH was adjusted to 7.5, its NMR spectrum demonstrated the undisputable presence of ATP, ADP and inorganic pyrophosphate, (PP_i in Figure 5, spectrum (d)). This spectrum shows an ATP/ADP ratio of about 0:6, which is typical for the granule adenine nucleotides (Table 1) and different from the whole (untreated) platelet extract ratio of about 1:6 (Figure 4).

The experiment shown in Figure 5 verifies that the NMR signals from ATP and ADP in intact platelets stem exclusively from the cytoplasmic nucleotides and that the granule nucleotides do not give measurable NMR signals in human platelets. PP_i is a major constituent of human platelet dense granules which is secreted from the cells upon thrombin activation.[13] The NMR resonances of human platelet granule nucleotides also became detectable after their release into the surrounding medium by treatment of the platelets with thrombin (see Figure 3 in Ugurbil *et al.*[35]). The spectra shown in Figures 4 and 5 were taken at 4°C. Raising the temperature to 37°C failed to give resonances from the granule nucleotides in intact human platelets (data not shown).

Studies on Intact Porcine Platelets, Their Extracts and Isolated Dense Granules

Spectra of untreated and $H_2O_2^-$-treated porcine platelets and of their $HClO_4$ extracts at 4°C were identical to those of human platelets (see Figures 4 and 5; Figure 6, trace (a)). However, when the spectra of the same platelet suspension were taken at 37°C, the resonances of the ATP-β phosphate and that of the terminal phosphates of ADP and ATP split into two peaks (Figure 6, spectrum (b)). Removal of the cytoplasmic adenine nucleotides by treating the platelets with 2-deoxyglucose + Antimycin A caused disappearance of the left peaks of the two double peaks (compare spectrum (b) with (c) in Figure 6). In this experiment the platelets had been prelabeled with ^{14}C-adenine, and the disappearance of the "left" peaks corresponded well with the disappearance of ^{14}C-ATP and ADP.[35] Hence, the "left" peaks of the ATP-β-phosphate and terminal phosphate resonances must stem from cytoplasmic ATP and ADP, while the 2-deoxyglucose/antimycin A-resistant "right" peaks are derived from the same phosphates of the granular ATP and ADP; their designations "cy" and "gr," respectively, are given in spectrum (b) of Figure 6. The pH inside dense granules is 5.5-5.8,[4, 8, 14, 22, 38,] and the β- and particularly the γ-phosphate of ATP shifts to higher ppm values on acidification from pH 7.3 to pH 5.5 (Figure 2). Thus the splitting of the two ATP peaks is due to pH differences between cytoplasm (pH 7.3) and the granules (pH 5.7). However, the granule pH cannot be read directly from the titration curve in Figure 2 (4.5 mM ATP) since the resonances are highly dependent on nucleotide concentration (Figure 3A) which is very high within the granules (Table 1).

The ^{31}P-NMR resonances from granule ATP and ADP in porcine platelets were detectable at 37°C but not at 4°C (compare (a) and (c) in Figure 6). A detailed study of temperature effects showed that the line widths increased as the temperature was lowered below 30°C, and the integrated intensities decreased

(Figure 7). The α-phosphate peak was the least affected and was even visible at 10°C. This 10.6 ppm position corresponded to the peak C in Figures 4 and 5, suggesting assignment of this peak to the α-resonances of the granule nucleotides in human platelets.

Isolated granules from porcine platelets also showed resonances that varied with temperature in exactly the same manner as intact platelets with their cytoplasmic nucleotides removed.[32] When the granule (ATP_γ + ADP_β)-resonances of normal porcine platelets (containing serotonin) were compared to platelets from a reserpine-treated pig (containing no serotonin), the temperature dependence of the band widths was slightly, but significantly changed (Figure 8A). A similar effect of serotonin on the temperature dependence was also found for the (ATP_γ + ADP_β)-resonances in the gel-like material forming from mixtures of ATP and ADP and Mg^{2+} (nucleotide:metal = 2) at nucleotide concentrations above 100 mM at 4°C and 250 mM at 37°C (Figure 8B). In both cases the slope of the *l*n (band width) versus the reciprocal absolute temperature lines were almost the same, but they were displaced from each other so that the bands were wider in the absence of amine at a given temperature. ^{31}P-NMR spectra of isolated

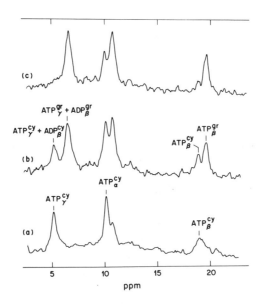

FIGURE 6. ATP and ADP resonances from porcine platelets at 145.7 MHz. Suspensions of porcine platelets were prepared according to Salganicoff and Fukami[27] and ^{31}P-NMR spectra were obtained from the same suspension (a) at 4°C (1000 scans), (b) at 37°C, 0.5-6 min after addition of 12 μg/ml of Antimycin A and 100 mM 2-deoxyglucose (500 scans) and (c) at 37°C, 9-14.5 min after addition of the inhibitors (500 scans). All spectra were obtained with 45° pulses and 0.68 sec repetition time. Superscripts "cy" and "gr" denote cytoplasm and granula, respectively.[35] (Originally published in *Proceedings* of National Academy of Sciences.)

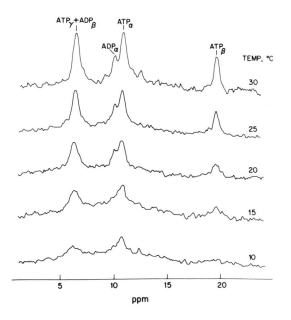

FIGURE 7. Temperature dependence of granule ATP and ADP resonances in porcine platelets. Each spectrum is the sum of 1000 scans obtained with 45° pulses and 0.68 repetition time. Platelets were incubated with 5 µg/ml Antimycin A and 30 mM 2-deoxyglucose in plasma before the preparation of the NMR sample, which also contained 50 mM 2-deoxyglucose.[35] (Originally published in *Proceedings* of National Academy of Sciences.)

porcine platelet granules also showed the same temperature dependence.[32] The peaks in the ^{1}H-NMR spectra from all dense granule constituents, including proteins, broadened and finally disappeared as the temperature decreased.[33] Thus, irrespective of which nucleus of the constituents' atoms was being monitored, i.e., ^{31}P (Figures 7 and 8), ^{19}F in 4,6-difluoroserotonin[7] or ^{1}H,[33] the NMR resonances broadened by lowering the temperature. These results demonstrate that interactions involving the nucleotides and the Mg^{2+} ions are predominant in determining the physicochemical state of the granule contents. Apparently, the nucleotides and Mg^{2+} form fluid aggregates that serve as a matrix for serotonin, and when the amine is incorporated into this matrix, its fluidity tends to increase. The ^{1}H-NMR measurements discussed below both confirm and expand this matrix model which differs from the only model previously proposed,[1, 2, 25] emphasizing the affinity between nucleotides and serotonin, but with no function of divalent cations.

Differences Among Species Regarding the NMR Resonances of Granule Constituents

In our studies the granule nucleotides gave measurable, temperature-dependent ^{31}P-NMR signals in porcine platelets, while those in human platelets did not.

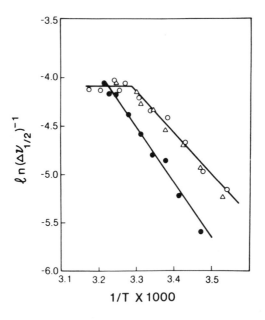

FIGURE 8A. Effect of temperature on band widths of granule nucleotide ^{31}P-resonances in porcine platelets or Mg^{2+}-nucleotide gels with or without serotonin. $Ln(\Delta V_{1/2})^{-1}$ (where $V_{1/2}$ = full width at half-maximum) of the $(ATP_\gamma + ADP_\beta)$ resonance have been plotted against T^{-1}. Granule nucleotides are charted from normal porcine platelets (o, Δ) and platelets from a shoat receiving injections of reserpine + atropine for altogether 4 days (●). The data were obtained with adenine nucleotide-depleted platelets as described in Figure 6. The straight line for the normal platelets was based on the (o) points.[32] (Reprinted with permission from *Biochemistry* 23, copyright 1984, American Chemical Society.)

These ^{31}P-NMR results have been confirmed by others who also demonstrated temperature-dependent granule nucleotide signals from granules of bovine and lapine platelets.[3, 29] Similar observations were made with ^{19}F-NMR in which 4,5-difluoroserotonin in the dense granules of intact porcine platelets gave temperature-dependent resonances, while that in the granules of human platelets gave no measurable ^{19}F-signals.[5, 6, 7]

The detectability of NMR resonances of granule nucleotides in animal platelets, but not of these nucleotides in human platelets, is believed to be due to the nature of the major divalent cation in the granules. Human platelet dense granules contain Ca^{2+} almost exclusively, while pig granules contain primarily Mg^{2+} and trace amounts of Ca^{2+} (Table 1). The dense granules from the other species studied with NMR (bovine, lapine) contain a 1:1 mixture of Mg^{2+} and Ca^{2+}.[24] As noted above, ADP and ATP form a gel with Mg^{2+} at high nucleotide concentrations. The nucleotides in these gels give ^{31}P-NMR spectra that are indistinguishable from those in intact dense granules[32] with their typical temperature dependence (Figures 8A and 8B). In contrast, ATP and ADP form an insoluble solid material with Ca^{2+}. The NMR signals, especially the band widths,

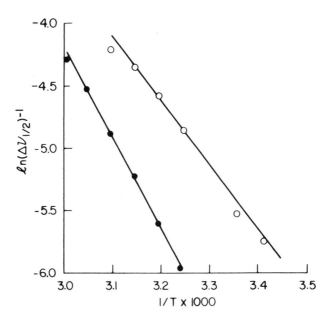

FIGURE 8B. The gel-like phase separating from mixtures of 0.4 M ATP + 0.4 M ADP + 1.6 M MgCl$_2$ at pH 5.3, with (●) and without (○) serotonin.[32] (Reprinted with permission from *Biochemistry* 23, copyright 1984, American Chemical Society.)

are very much broadened when the rotational motion of a molecule is restricted, such as in the solid state,[34] an effect that fully explains the undetectability (i.e., broadening) of the NMR resonances from ATP, ADP, serotonin and pyrophosphate in the Ca^{2+}-dominated human platelet dense granules.

^{1}H-NMR OF PORCINE PLATELET DENSE GRANULE CONSTITUENTS

Spectra of Organelles

^{1}H-NMR spectra (aromatic hydrogens) of cytoplasmic organelles (dense granules, α-granules, mitochondria, lysosomes) and purified dense granules isolated by sucrose density gradient centrifugation[28] are shown in Figure 9 with assignments of the peak resonances as explained elsewhere.[33] The granule preparations had been equilibrated with D$_2$O in order to replace the exchangeable ^{1}H atoms (amine-H, phosphate-H, hydroxyl-H) by D (^{2}H) in order to reduce the number of ^{1}H resonances in the spectra. The major (^{1}H)-peaks 2 and 3 originate, respectively, from the hydrogens in the 8- and 2-positions of the purine rings, while peak 8 originates from the hydrogen (H-1') on the ribose moiety (Figure 9(a)). We will see below that the resonances from just these hydrogens are instrumental for the construction of the molecular arrangement of the nucleotide stacking within the dense granules.

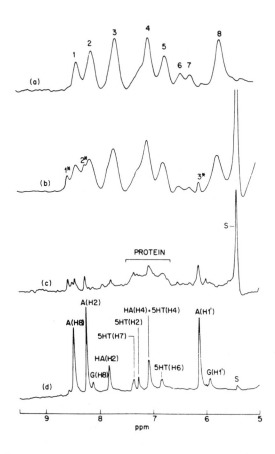

FIGURE 9. The ^{1}H-NMR spectra (360 MHz, 35°C) of porcine platelet granule suspensions and their extract. (a) is from a suspension of cytoplasmic organelles, i.e., dense granules, α-granules, mitochondria, lysosomes (400 FIDs obtained with 60° pulses, 2.2 sec repetition times). (b) is from a suspension of purified dense granules (200 FIDs, 45° pulses, 3 sec repetition times). (c) is the supernatant (pH 7.2) after removal of the dense granules in the suspension shown in (b) by centrifugation (100 FIDs, 45° pulses, 5.5 sec repetition times). (d) is a perchloric acid extract of the granules isolated by centrifugation of the dense granule suspension in (b). A and G represent adenine and guanine nucleotides, respectively. Peak S is sucrose. See text for further explanation.[33] (Reprinted with permission from *Biochemistry* 23, copyright 1984, American Chemical Society.)

The smaller peaks, no. 5, 6 and 7 from the cytoplasmic organelles (Figure 9(a)) originate from the hydrogens in serotonin's indole ring, i.e., positions 7 + 2, 4 and 6, respectively, and peak 1 is probably from H-2 in histamine. In the well-resolved NMR spectrum from an acid extract of purified dense granules (Figure 9(d)) the assignments are shown. Small resonances from H in the 8 position of the purine ring and of the 1'-position in the ribose of guanine nucleotides and a distinct resonance from H-2 of histamine are also present (Figure 9(d)). The chemical shifts of H-8 (peak 2) and H-2 (peak 3) of the adenine nucleotides and particularly of H-2 in histamine in the cytoplasmic organelles (Figure 9(a)), are different from the resonances of the corresponding hydrogens in the extract (Figure 9(d)). This is because the pH differs in the intact granules (pH = 5.7) and the extract (pH = 7.0). The great pH dependence of histamine's H-2 will be used below as a sensitive pH monitor of intergranular pH during serotonin accumulation.

Since the pH of both the granule interior and the suspending medium measure 5.7 and 7.3, respectively, the pH-sensitivity of the chemical shift in some of the hydrogens discussed above allows determination of the concentration of intra- and extragranular substances directly in the granule suspension. This is shown in Figure 9(b) where the resonance peaks from extragranular adenine nucleotides (peaks 2* and 3*) and histamine (peak 1*) are marked with asterisks. The validity for assigning these extragranular resonances can be seen from the supernant spectrum after pelleting the granules from the suspension (Figure 9(c)). A comparison of the ^{1}H-NMR spectra of cytoplasmic organelles (Figure 9(a)) and of purified dense granules (Figure 9(b)) shows that the organelle suspension does not contain demonstrable extragranular, granule-derived substances, while the purified dense granule suspension does. As shown elsewhere,[33] the amount of extragranular substances increased rapidly (threefold in 30 min) in the suspensions of purified dense granules at 37°C in the NMR tube. Thus, the granules in these suspensions were unstable and lysed progressively, and suspensions of the stable cytoplasmic organelles were used in subsequent experiments. Figure 9 shows, finally, that a large, broad resonance (peak 4) did not stem from the dense granule nucleotides and amines, but rather from a number of protein hydrogens (peak 4 is absent in Figure 9(d) where proteins had been removed by acid precipitation). These proteins are a mixture of extragranular proteins (added albumin, Figure 9(c)) and granule-bound proteins.

Uptake of Serotonin

The effect on the ^{1}H-NMR spectrum when serotonin was added to a suspension of cytoplasmic organelles is shown in Figure 10. Two important differences can be seen by comparing the spectrum of granules to which serotonin was added (Figure 10(b)) with the control spectrum (Figure 10(a)): 1) serotonin peaks 6 and 7 have increased intensity due to transport of the amine into the granules; 2) the pH-sensitive resonance of histamine (peak 1) has moved to lower ppm values due to alkalinization of the granule interior. Uptake of serotonin did not, however, cause any changes in size or chemical shift of the nucleotide resonances (peaks 2, 3 and 8, Figure 10). The ppm values of intragranular serotonin resonances (peaks 6 and 7) were lower than the extragranular serotonin values (marked 5HT in Figure 10(b)) due, most likely, to the association with adenine nucleotides, as demonstrated above (Figure 3B). The alkalinization of the granule interior upon serotonin entry suggests that serotonin uptake is coupled to the proton gradient (acidic inside) across the granule membrane, and thus, agrees with other studies.[4, 22, 26, 38] Serotonin may simply be transported as the neutral species across the granule membrane and remain trapped inside the granule in the protonated form.

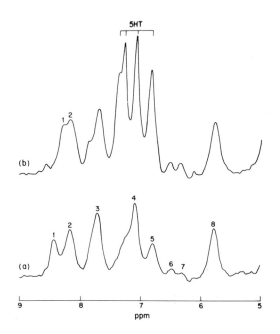

FIGURE 10. Effect of added serotonin on ^{1}H resonances from dense granules. The spectra are from a suspension of cytoplasmic organelles at 35°C (600 FIDs, 90° pulses, 3 sec repetition time, convolution difference procedure). (a) is a control spectrum of the organelles before serotonin addition, and (b) is a spectrum after addition of 10 mM serotonin. The four resonances labeled 5HT stem from extragranular serotonin. (Reprinted with permission from *Biochemistry* 23, copyright 1984, American Chemical Society.)

Nuclear Overhauser Effects Between Adjacent Nucleotides and Between Nucleotide and Serotonin in Dense Granules

Energy saturation in one proton will initiate energy transfer to neighboring protons less than 4 Å from the saturated proton. This pumping of energy into a spin reduces the proton's resonance intensity. When the ribose hydrogen in 1-position was saturated in a suspension of dense granules, the intensities of peak 2 (H-8 of adenine) and peak 3 (H-2 of adenine) were increasingly reduced as the time of saturation of ribose H-1 was prolonged (Figure 11). Some effects of ribose H-1 saturation were observed in the other ribose hydrogens showing resonances to the right of the DHO peak (Figure 11). This was expected since these hydrogens are within 4 Å of each other.

However, the strong Nuclear Overhauser effects (NOE) between ribose H-1' and adenine H-2 and H-8 give important information about the molecular arrangement of the nucleotides within the granules. As discussed in detail elsewhere,[31, 33] the energy transfer shown cannot occur between H-1' and H-2/H-8 on the same molecule, but between the ribose H-1' on one adenine nucleotide and the adenine H-2 and H-8 in an adjacent nucleotide. From the fractional reduction in intensity and irradiation times, one can calculate that the distance

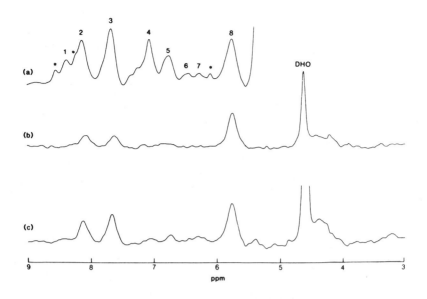

FIGURE 11. Nuclear Overhauser effects (NOE) on ^{1}H resonances from dense granules upon saturation of peak 8. (a) control spectrum. (b) NOE difference spectrum with 0.15 sec saturation of H-1' (peak 8). (c) NOE difference spectrum with 0.5 sec saturation of H-1'. (For further details, see Ugurbil *et al.*[33] Reprinted with permission from *Biochemistry* 23, copyright 1984, American Chemical Society.)

between the ribose H-1' on another molecule and the adenine H-8 and H-2 is 2.7 and 2.8 Å, respectively.[33] This provides direct evidence of the nucleotides' association with one another within the granules, and it is probably the reason for the upfield shift of the adenine protons in the granules compared to the resonances from nucleotides free in solution (Figure 9).

Tentative Molecular Model of Nucleotide Aggregate

Considering the distances between ribose H-1 and adenine H-2 and H-8 in a neighboring ATP molecule (above), and based on the assumptions that ribose and adenine rings of each molecule are in the *syn*-configuration[33] and that a Mg/ nucleotide ratio of 2 exists within the granules (Table 1), we have constructed a molecular model showing possible arrangement of Mg and ATP within the porcine dense granules (Figure 12A). The plane adenine rings are stacked in parallel over each other, but they do not overlap. Instead, if one looks down the axis perpendicular to the plane of the adenine rings, each ring is skewed to the right of the ring just beneath it, i.e., they form a right-handed spiral or helix, moving upward. We have assumed that the individual ATP molecules are held together with two ions of Mg^{2+} in such a way that both charges on the Mg ion are neutralized with one negative charge from a phosphate ion in one ATP molecule, and one from a phosphate ion in the neighboring ATP molecule. This linking of the triphosphate moieties is not based on the NMR studies, but it is thought to be

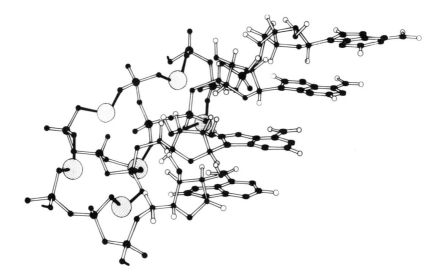

FIGURE 12A. Model of the ATP-Mg^{2+} complex present in porcine dense granules based on the ^{1}H-NMR data and other observations in this paper. Four ATP molecules in ball and stick models have been stacked on top of each other with their adenine rings (right) in parallel planes and the ribose H-1 placed 2.7-2.9 Å from the H-8 and H-2 on the adjacent adenine ring. The ribose rings are positioned in *syn*-configuration (about 30° to the adenine plane). Each ATP molecule is linked noncovalently (black pins) to the one above and the one below with two Mg^{2+} ions (large, dotted balls) between the O$^-$ of the γ- and β-phosphoryls of adjacent triphosphate moieties on the left. Atoms: H = white balls; O, C and N = black balls or disks (aromatic rings), P = black balls, larger than the other atoms.

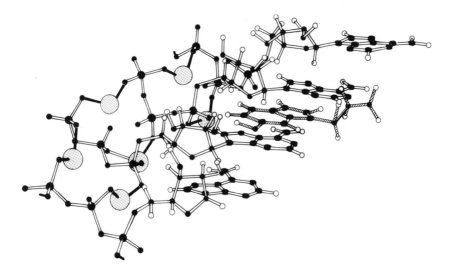

the most likely molecular arrangement of stacked ATP molecules and Mg^{2+} in a 1:2 ratio. Other possible arrangements of Mg^{2+} and stacked ATP should be possible, although they would have to involve interactions between the negatively charged triphosphate moieties on two adjacent ATP molecules and the positively charged Mg ion. Thus, except from the intermolecular "Mg-phosphate bonding," the similarity between our model and poly-AMP or any DNA molecule is striking. We have estimated[33] that the size of the Mg_2ATP aggregates is in the order of 10-15 Kd, equivalent to 20-30 molecules of nucleotide per aggregate. In the 4-nucleotide piece of an aggregate, shown in Figure 12A, the ATP molecule turns about 90° from the lowest to the highest. In a 20-30 nucleotide aggregate, the nucleotide chain thus makes a spiral of 1.25-1.88 turns.

When the H-4 hydrogen on serotonin was saturated, reduction in the intensities of the 2-, 8- and 1'-hydrogens of the nucleotides took place, although these NOEs were less than those between H-1' and H-2/H-8.[33] This suggests that serotonin is present close to the nucleoside part of the nucleotides in the aggregates. It is possible in our model that serotonin intercalates between adjacent nucleoside moieties in the aggregate, as shown in Figure 12B, in a way that does not alter the structure of the nucleotide-Mg aggregate. The latter requirement is made because serotonin has little or no effect on the temperature dependence of the band widths shown in Figures 8A and 8B.

CONCLUSIONS

The large concentrations of ATP (443 mM) and ADP (255 mM) in porcine platelet dense granules exist as helical aggregates of 20-30 stacked nucleotides each, with a size of approximately 10-15 Kd, held together by Mg ions with a metal:nucleotide ratio of 2:1. The existence of such aggregates reduces the effective (solute) concentration of nucleotides to 17-25 mM and, hence, explains the osmotic stability of the dense granules within the cytoplasm. Serotonin is intercalated between adjacent purine rings in the nucleotide-Mg matrix without appreciably altering the physical state of the aggregates. The amine is transported across the granule membrane in the neutral form and trapped within the granules in the protonated state followed by intercalation. The physical state of the nucleotide-Mg aggregates is temperature-dependent, converting into immobile complexes as the temperature is lowered. This temperature dependence, as well as the difference in pH between granule interior (pH 5.7) and cytoplasm (pH 7.3), directly distinguishes the granule and cytoplasmic pools of adenine nucleotides in NMR spectra in suspensions of porcine platelets.

In human platelets, $(^{31}P)NMR$ also distinguishes between granule and cytoplasmic nucleotide pools, but the granule nucleotides are immobilized to an extent that prevents studies with NMR. This is probably because they exist with Ca^{2+} as solid state complexes. These (putative) nucleotide-Ca aggregates may act as matrixes into which serotonin is intercalated in the same manner as the (porcine) helical nucleotide-Mg stacks.

ACKNOWLEDGMENTS

The help of Sissel Rongved and Rolf Isrenn for building the molecular model of the ATP aggregate and the drawing of this model by Liv Skarstein is highly appreciated.

REFERENCES

1. **Berneis, K.H., Da Prada, M., and Pletscher, A.**, Micelle formation between 5-hydroxytryptamine and adenosine triphosphate in platelet storage organelles, *Science*, 165, 913, 1969.
2. **Berneis, K.H., Pletscher, A., and Da Prada, M.**, Metal-dependent aggregation of biogenic amines: A hypothesis for their storage and release, *Nature*, 224, 281-283, 1969.
3. **Carroll, R.C., Edelheit, E.B., and Schmidt, P.G.**, Phosphorus nuclear magnetic resonance of bovine platelets, *Biochemistry*, 19, 3861-3867, 1980.
4. **Carty, S.E., Johnson, R.G., and Scarpa, A.**, Serotonin transport in isolated platelet granules. Coupling to the electrochemical proton gradient, *J. Biol. Chem.*, 256, 11244-11250, 1981.
5. **Costa, J.L., Dobson, C.M., Fay, D.D., Kirk, K.L., Paulsen, F.M., Valerie, C.R., and Vecchione, J.J.**, Nuclear magnetic studies of amine storage in pig platelets, *FEBS Lett.*, 136, 325-328, 1981.
6. **Costa, J.L., Dobson, C.M., Kirk, K.L., Paulsen, F.M., Valerie, C.R., and Vecchione, J.J.**, Studies on human platelets by ^{19}F and ^{31}P NMR, *FEBS Lett.*, 99, 141-146, 1979.
7. **Costa, J.L., Dobson, C.M., Kirk, K.L., Paulsen, F.M., Valerie, C.R., and Vecchione, J.J.**, Nuclear magnetic resonance studies of blood platelets, *Philos. Trans. R. Soc. Lond. [Biol.]*, 289, 423-427, 1980.
8. **Costa, J.L., Eanes, E.D., Fay, D.D., and Hailer, A.W.**, Preparation and characterization of synthetic models for the dense granules of human platelets, *Cell Calcium*, 2, 459-472, 1981.
9. **Costa, J.L., Reese, T.S., and Murphy, D.L.**, Serotonin storage in platelets: Estimation of storage-package size, *Science*, 183, 537-538, 1974.
10. **Daniel, J.L., Molish, I.R., and Holmsen, H.**, Radio-labeling of purine nucleotide pools as a method to distinguish among intracellular compartments: Studies on human platelets, *Biochim. Biophys. Acta*, 632, 444-453, 1980.
11. **Daniel, J.L., Molish, I.R., Robkin, L., and Holmsen, H.**, Nucleotide exchange between cytosolic ATP and F-actin-bound ADP may be a major energy-utilizing process in unstimulated platelets, *Eur. J. Biochem.*, 156, 677-684, 1986.
12. **Daniel, J.L., Robkin, L., Molish, I.R., and Holmsen, H.**, Determination of the ADP concentration available to participate in energy metabolism in an actin-rich cell, the platelet, *J. Biol. Chem.*, 254, 7870-7873, 1979.
13. **Fukami, M.F., Dangelmaier, C.A., Bauer, J., and Holmsen, H.**, Secretion, subcellular localization and metabolic status of inorganic pyrophosphate in human platelets: A major constituent of the amine-storing granules, *Biochem. J.*, 192, 99-105, 1977.
14. **Grinstein, S., and Furuaya, W.**, Intracellular distribution of acridine derivatives in platelets and their suitability for cytoplasmic pH measurements, *Biochim. Biophys. Acta*, 803, 221-228, 1985.
15. **Haslam, R.J.**, Role of adenosine diphosphate in the aggregation of human blood-platelets by thrombin and by fatty acids, *Nature*, 202, 765-766, 1964.
16. **Holmsen, H.**, Nucleotide metabolism of platelets, *Annu. Rev. Physiol.*, 47, 677-690, 1985.
17. **Holmsen, H.**, Platelet secretion, in *Hemostasis and Thrombosis—Basic Principles and Clinical Practice*, 2nd edition, Colman, R.W., Hirsh, J., Marder, V.J., and Salzman, E.W., Eds., Lippincott, Philadelphia, 1987, 606-617.

18. **Holmsen, H., and Robkin, L.,** Hydrogen peroxide lowers ATP levels in platelets without altering adenylate energy charge and platelet function, *J. Biol. Chem.,* 252, 1752-1757, 1977.

19. **Holmsen, H., and Robkin, L.,** Effects of antimycin A and 2-deoxyglucose on energy metabolism in washed human platelets, *Thromb. Haemost.,* 42, 1460-1472, 1980.

20. **Holmsen, H., Robkin, L., and Driver, H.A.,** Reduction of the metabolic adenylate pool in platelets by hydrogen peroxide: Studies on the mechanism, *Fed. Proc.,* 38, 826, 1979.

21. **Holmsen, H., and Weiss, H.J.,** Secretable pools in platelets, *Annu. Rev. Med.,* 30, 119-134, 1979.

22. **Johnson, R.G., Scarpa, A., and Salganicoff, L.,** The internal pH of isolated serotonin-containing granules of pig platelets, *J. Biol. Chem.,* 253, 7061-7068, 1978.

23. **Kaplan, K.,** In vitro platelet responses: α-granule secretion, in *Platelet Responses and Metabolism,* Vol. 1, Holmsen, H., Ed., CRC Press, Boca Raton, 1987, 145-162.

24. **Meyers, K.M., Holmsen, H., and Seachord, C.L.,** Comparative study of platelet dense granule constituents, *Am. J. Physiol.,* 243, R451-R461, 1982.

25. **Pletscher, A., Da Prada, M., and Berneis, K.H.,** Aggregation of biogenic monoamines and nucleotides in subcellular storage organelles, *Mem. Soc. Endocrinol.,* 19, 767-770, 1971.

26. **Rudnick, G., Fishkes, H., Nelson, P., and Schuldiner, S.,** Evidence for two distinct serotonin transport systems in platelets, *J. Biol. Chem.,* 255, 3638-3640, 1980.

27. **Salganicoff, L., and Fukami, M.H.,** Energy metabolism of blood platelets. Isolation and properties of platelet mitochondria, *Arch. Biochem. Biophys.,* 153, 726-735, 1972.

28. **Salganicoff, L., Hebda, P.A., Yandrasitz, J., and Fukami, M.H.,** Subcellular fractionation of pig platelets, *Biochim. Biophys. Acta,* 385, 394-411, 1975.

29. **Schmidt, P.G., and Carroll, R.C.,** [31]P Nuclear resonance studies of platelet dense granule nucleotides, *Biochim. Biophys. Acta,* 715, 240-245, 1982.

30. **Sixma, J.J.,** Morphology, in *Platelet Responses and Metabolism,* Vol. 1, Holmsen, H., Ed., CRC Press, Boca Raton, 1987, 33-62.

31. **Ugurbil, K.,** Studies on storage mechanisms in the dense granules: Nuclear magnetic resonance, in *Platelet Responses and Metabolism,* Vol. II, Holmsen, H., Ed., CRC Press, Boca Raton, 1987, 153-170.

32. **Ugurbil, K., Fukami, M.H., and Holmsen, H.,** ^{31}P-NMR studies of nucleotide storage in the dense granules of pig platelets, *Biochemistry,* 23, 409-416, 1984.

33. **Ugurbil, K., Fukami, M.H., and Holmsen, H.,** Proton NMR studies of nucleotide and amine storage in the dense granules of pig platelets, *Biochemistry,* 23, 416-428, 1984.

34. **Ugurbil, K., and Holmsen, H.,** Nucleotide compartmentation: Radioisotopic and nuclear magnetic resonance studies, in *Platelets in Biology and Pathology 2,* Gordon, J., Ed., Elsevier/North Holland Biomedical Press, Amsterdam, 1981, 147-177.

35. **Ugurbil, K., Holmsen, H., and Schulman, R.G.,** Adenine nucleotide storage and secretion in platelets studied by ^{31}P nuclear magnetic resonance, *Proc. Natl. Acad. Sci. USA,* 76, 2227-2231, 1979.

36. **Van Oost, B.A.,** In vitro platelet responses: Acid hydrolase secretion, in *Platelet Responses and Metabolism,* Vol. I, Holmsen, H., Ed., CRC Press, Boca Raton, 1987, 163-192.

37. **Verhoeven, A.J.M., Mommersteeg, M.E., and Akkerman, J.W.N.,** Balanced contribution of glycolytic and adenylate pool in supply of metabolic energy in platelets, *J. Biol. Chem.,* 260, 2621-2624, 1985.

38. **Wilkins, J.A., and Salganicoff, L.,** Participation of a transmembrane proton gradient in 5-hydroxytryptamine transport by platelet dense granules and dense granule ghosts, *Biochem. J.,* 198, 113-123, 1981.

Chapter 4

THE ROLE OF INOSITOL PHOSPHATE METABOLISM IN PLATELET STIMULUS-RESPONSE COUPLING

James L. Daniel

INTRODUCTION

This chapter will focus on one aspect of stimulus-response coupling in platelets, namely the role of inositol phosphate metabolism in platelet activation. For a more comprehensive review on platelet stimulus-response coupling see Huang and Detwiler[43] or Haslam.[40] For a more complete discussion of the metabolism of phosphatidylinositol, see chapter 5 in this book.

INOSITOL PHOSPHATE METABOLISM AND PLATELET ACTIVATION

Phosphatidylinositol (PtdIns) Turnover

Inositol-containing phospholipids are minor components of the membranes of most mammalian cells, including platelets, where the inositol is found in about 5% of total phospholipid.[18,69] The fatty acid composition of PtdIns is somewhat unique in that over 80% consists of the molecular species that contains stearic acid in the 1 position and arachidonic acid in the 2 position.[70] In addition to PtdIns, mammalian cells contain smaller amounts of the polyphosphoinositides, phosphatidylinositol 4-monophosphate (PtdIns 4-P), and phosphatidylinositol 4,5-bisphosphate (PtdIns 4,5-P_2).[100] The two polyphosphoinositides are present at 10% of the amount of PtdIns and are formed by the reversible phosphorylation of PtdIns by ATP, probably by a specific set of kinases and phosphatases. The interconversion of the members of the phosphoinositide family in human platelets is very rapid as judged by the fact that the phosphates in 4 and 5 positions attain the same specific radioactivity as metabolic ATP in a very short time.[25,99,100]

Interest in PtdIns metabolism was generated when early experiments by Hokin and Hokin[41] showed that hormonal stimulation of a receptor could lead to enhanced incorporation of [^{32}P]-PO_4 into PtdIns. Subsequent studies on a wide variety of cells demonstrated that hormonal stimulation can often lead to increased metabolism of PtdIns.[9] Platelets show this type of response when stimulated by thrombin[64] and other agonists.[65,67] The enhanced incorporation of [^{32}P]-PO_4 into PtdIns was caused by increased turnover of the so-called "PI cycle" which was initiated by phosphodiesteratic cleavage of PtdIns by phospholipase C (PLC) (see Figure 1, reactions 1-4). In 1975, Michell[72] hypoth-

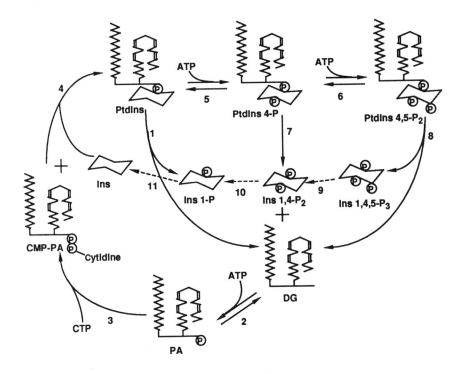

FIGURE 1. Pathways breakdown and resynthesis of phosphatidylinositol (PtdIns) and phosphatidylinositol polyphosphates.

esized that a PLC-catalyzed breakdown of PtdIns was an important step in hormonal reactions that lead to mobilization of intracellular stores of calcium. While further support for this hypothesis was provided by the studies of Michell and Kirk,[73] there was little information about the biochemical steps in between the activation of PLC and release of internally stored Ca^{2+}.

A key breakthrough in understanding the relevance of the "PI response" to cell activation came with the recognition that the initial event that occurred when liver cells were stimulated was the hydrolysis of PtdIns 4,5-P_2.[58,74] Thus, a measured loss in the mass of PtdIns could result from hydrolysis of PtdIns 4,5-P_2 and subsequent resynthesis of PtdIns 4,5-P_2 via phosphorylation of PtdIns. This observation has been confirmed in other cells, including platelets;[2,12,70] however, since interconversion of the phosphoinositides is rapid, a decrease in the amount of PtdIns 4,5-P_2 is difficult to detect. More direct evidence for a PLC-catalyzed hydrolysis of PtdIns 4,5-P_2 came from the demonstration that the reaction product, inositol 1,4,5-trisphosphate (Ins 1,4,5-P_3), is produced more rapidly than inositol monophosphate.[8] These observations lead to the investigation of the Ins 1,4,5-P_3 function and ultimately to an explosion of research in this area.

The Second Messenger Function of Ins 1,4,5-P$_3$

The fact that Ins 1,4,5-P$_3$ was the first water-soluble molecule released when cells were hormonally stimulated suggested that it might be the missing link between the events occurring at the cell membrane and the release of Ca^{2+} from intracellular stores. The obvious experiment is to add Ins 1,4,5-P$_3$ to a cell in the absence of extracellular Ca^{2+} to determine if intracellular Ca^{2+} increases; however, since the Ins 1,4,5-P$_3$ site of action is presumed to be at an intracellular membrane, probably the endoplasmic reticulum, and since Ins 1,4,5-P$_3$ is highly charged, this experiment is hampered by an inability to deliver Ins 1,4,5-P$_3$ to its site of action. It was found that pancreatic acinar cells could be partially permeabilized if incubated in a medium containing very low Ca^{2+} and ATP.[93] When Ins 1,4,5-P$_3$ was added to these cells, an increase in Ca^{2+} in the medium was detected using a Ca^{2+}-sensitive electrode. Half maximal release of Ca^{2+} was achieved at less than 1 µM Ins 1,4,5-P$_3$, and the released Ca^{2+} did not appear to be derived from the mitochondria. These basic observations were immediately confirmed for the case of hepatocytes.[16,56] The release of Ca^{2+} by Ins 1,4,5-P$_3$ was specific in that sugar diphosphates and inositol mono- and diphosphates had little or no effect on Ca^{2+} release.[10] Recent studies of Strupish et al.[94] have shown the release of Ca^{2+} mediated by Ins 1,4,5-P$_3$ is very stereospecific. The L-Ins 1,4,5-P$_3$ isomer was over 1000 times less potent in releasing Ca^{2+} from permeabilized GH$_3$ and Swiss 3T3 cells than was the D isomer.

In platelets, the first evidence for a potential second messenger function for Ins 1,4,5-P$_3$ came from O'Rourke et al.,[77] who isolated a fraction of membrane vesicles that accumulated Ca^{2+} in the presence of ATP. Ins 1,4,5-P$_3$ caused a rapid and dose-dependent release of the majority of the sequestered Ca^{2+}. Two similar studies appeared at about the same time, the first by Authi and Crawford,[3] who prepared intracellular membranes by free flow electrophoresis and showed that Ins 1,4,5-P$_3$ induced release of stored Ca^{2+} from this preparation with an EC$_{50}$ of 0.3-0.6 µM. Adunyah and Dean[1] also confirmed this observation and determined that half-maximal release occurred at 0.4 µM Ins 1,4,5-P$_3$. Brass and Joseph[13] showed that saponin-permeabilized platelets would sequester ^{45}Ca^{2+} into both mitochondrial and nonmitochondrial internal stores. Ins 1,4,5-P$_3$, when added to the medium, caused a release of ^{45}Ca^{2+} from the nonmitochondrial store with an EC$_{50}$ of 0.8 µM. The Ins 1,4,5-P$_3$-induced ^{45}Ca^{2+} release was potentiated by an increase in pH. Ins 1,4,5-P$_3$ caused release of granule-stored serotonin which was attributed to an effect of the released Ca^{2+} rather than as a direct effect of Ins 1,4,5-P$_3$.

Several investigators have demonstrated that introduction of Ins 1,4,5-P$_3$ into permeabilized platelets can have effects beyond simple elevation of Ca^{2+}. Lapetina et al.[61] demonstrated that Ins 1,4,5-P$_3$ could enhance protein phosphorylation in saponin-permeabilized platelets. Israels et al.[53] found that saponinized platelets, when treated with Ins 1,4,5-P$_3$, underwent a variety of reactions similar to those induced by agonists. These reactions included aggregation, secretion, protein phosphorylation, arachidonic acid liberation and thromboxane formation. Pretreatment of the cells with aspirin blocked most of these responses suggesting that the primary action of Ins 1,4,5-P$_3$ was to liberate Ca^{2+} and thus activate phospholipase A2. The other effects were due to platelet activation by thromboxane A2 that would have been generated. Similar conclusions were drawn by other investigators.[4,14,107]

O'Rourke *et al.*[78] demonstrated that in a platelet membrane system containing both plasma and intracellular membranes, thrombin in the presence of GTP could cause release of ^{45}Ca^{2+} from the intracellular vesicles. The release of ^{45}Ca^{2+} was attributed to the formation of a soluble factor that could be destroyed by alkaline phosphatase. A monoclonal antibody was found that blocked the thrombin-induced ^{45}Ca^{2+} release in this system and also blocked Ins 1,4,5-P$_3$ induced Ca^{2+} release. These data suggested that the soluble factor that caused Ca^{2+} release was Ins 1,4,5-P$_3$ and that the monoclonal antibody blocked either the Ins 1,4,5-P$_3$ receptor or the channel whose opening it controlled. A 260,000 Da protein that has Ins 1,4,5-P$_3$ receptor properties has been purified from rat brain.[95]

Measurement of the Formation of Inositol Trisphosphate (InsP$_3$) in Platelets

Early Studies Employing [^{32}P]-PO$_4$-labeling

The first reported measurement of InsP$_3$ in human platelets was described by Agranoff *et al.*[2] In these experiments, human platelets were labeled with [^{32}P]-PO$_4$ and InsP$_3$ separated by paper electrophoresis in oxalate buffer. However, two aspects of these experiments were unsatisfying. The level of InsP$_3$ in resting cells appeared to be rather high. While no mass measurement was made, comparison with the radioactivity present in ATP (7 μM cytosolic concentration in resting platelets)[29] suggested that the resting concentration of InsP$_3$ might be about two orders of magnitude above the EC$_{50}$ for Ca^{2+} release discussed above. In addition, the change induced by thrombin in InsP$_3$ was rather disappointing, less than twofold above the control. While many [^{32}P]-labeled metabolites had been identified in platelets,[42] the system of Agranoff *et al.*[2] resolved only a few, suggesting that each radioactive spot might contain several metabolites (see below).

Studies Employing [^{3}H]-inositol-labeling

The other approach that has been used to measure agonist-mediated Ins 1,4,5-P$_3$ production in platelets employs [^{3}H]-inositol which is incorporated into PtdIns and polyphosphatidylinositols. This method's advantage is that only inositol-containing metabolites are labeled and thus the interference from other phosphate-containing compounds is not a problem. However, for studies in platelets the method is not without its disadvantages. The main problem is that [^{3}H]-inositol is incorporated into platelet lipids at a very slow rate, probably due to inadequate transport. It has not yet been demonstrated that all inositol-containing phospholipids are labeled uniformly. Some laboratories have used large amount of [^{3}H]-inositol in order to obtain sufficient labeling; however, the cost of [^{3}H]-inositol limits the number of experiments that can be done in this way. Increasing the [^{3}H]-inositol labeling time can lead to platelets that are either refractory or less responsive to weak platelet agonists. Another approach has been to include Mn^{2+} during labeling with [^{3}H]-inositol since this divalent ion enhances the rate of inositol uptake. However, Labarca et al.[60] have suggested that the [^{3}H]-inositol labeling that occurs is non-uniform and that receptor-sensitive pools of PtdIns are poorly labeled.

Several reports of agonist-induced release of InsP$_3$ using [^{3}H]-inositol-labeled platelets appeared almost at the same time, the first coming from Watson et al.,[104] who used rather large amounts of [^{3}H]-inositol and a 3 hr incubation in plasma-free buffer to obtain sufficient labeling. HPLC was used to separate inositol phosphates after extraction. They found that detectable increases in InsP$_3$ and InsP$_2$ could be seen after a 5 sec stimulation with thrombin. When Li$^+$ (which is thought to inhibit some mono- and polyphosphoinositol phosphatases) was included in the incubation buffer, increases in InsP were detected. The increase in InsP occurred about 15 sec after that of InsP$_3$ and InsP$_2$. In spite of the amount of [^{3}H]-inositol used and the time of incubation, the radioactivity in [^{3}H]-InsP$_3$ was only about 100 cpm above background. These studies were also confirmed by Siess and Binder[89] in human platelets. In their experiments, MnCl$_2$ was used to enhance [^{3}H]-inositol uptake. The increase in InsP$_3$ induced by thrombin was only about 50 cpm above control; however, much larger changes in inositol 1,4-bisphosphate (Ins(1,4)P$_2$) were seen. The major finding of this study was that InsP formation lagged behind the formation of the inositol polyphosphates. Vickers et al.[101] performed similar experiments using rabbit platelets incubated with 20 mM Li$^+$ and stimulated with thrombin. They found larger increases in [^{3}H]-InsP$_3$ than had been found for human platelets. Since more [^{3}H]-InsP$_3$ was formed, it appears that better labeling of inositol phospholipids is achieved in washed rabbit platelets than in human platelets.

Having established that InsP$_3$ was formed in platelets stimulated with thrombin, investigators were interested in determining whether other platelet agonists could produce similar inositol phosphate metabolism and, if they did, whether the changes might be qualitatively different. Watson et al.[106] measured InsP$_3$ formation in human platelets treated with collagen. Consistent with the lag seen in both ATP secretion and aggregation, formation of InsP$_3$ and InsP$_2$ was slower than in thrombin-stimulated platelets. InsP formation was clearly slower than the inositol polyphosphates. Addition of indomethacin blocked most of the aggregation and secretion and blunted, but did not abolish, InsP$_3$ and InsP$_2$ formation. However, under these conditions no change in intracellular Ca^{2+}, as measured by quin2 fluorescence, was observed. Interestingly, quin2-loaded platelets showed diminished secretion and aggregation, as might be expected from the Ca^{2+} buffering activity of quin2; however, in the dye-loaded cells, collagen-induced inositol polyphosphate formation was not changed. This experiment is consistent with the theory that $InsP_3$ causes liberation, rather than the converse. A stable endoperoxide analog, U44069, was studied and shown to induce rapid inositol polyphosphate formation consistent with the idea that indomethacin could reduce $InsP_3$ formation by reducing formation of stimulatory prostaglandins and thromboxanes. Several possible explanations for the uncoupling of $InsP_3$ formation and Ca^{2+} formation were presented.

Shukla[88] studied rabbit platelets prelabeled with $[^3H]$-inositol and challenged with 1×10^{-9} M platelet-activating factor (PAF). A similar pattern of increased inositol phosphate formation was seen as described for other agonists. $InsP_3$ formation was fivefold above background 5 sec after addition of PAF and was inhibited by forskolin, a stimulator adenylate cyclase. Studies from our laboratory have shown that PAF can also cause formation of $InsP_3$ in human platelets (Daniel *et al.*, unpublished data). Vasopressin[90] and serotonin[32] have been shown to cause production of $InsP_3$; however, only in the former study was feedback from cyclooxygenase products eliminated. Epinephrine-induced $InsP_3$ formation has not been demonstrated. However, it is unlikely that epinephrine causes a major stimulation since it does not produce a detectable change in intracellular Ca^{2+} nor induce formation of phosphatidic acid (PA).[68] The effect of the platelet agonist ADP on inositol phosphate metabolism is controversial and will be discussed below.

The Measurement of the InsP₃ Mass in Resting and Stimulated Platelets

As has been discussed, the concentration of Ins $1,4,5$-P_3 needed to release Ca^{2+} from intracellular pools was found to be in the range of several hundred micromolar Ins $1,4,5$-P_3. Thus, it would be expected that the level of $InsP_3$ in resting cells would be below this value and would rise above this value when platelets are treated with agonists that mobilize intracellular Ca^{2+}. Mass determinations were not possible from previous radiolabeling experiments since it was not known whether the radiolabel was in equilibrium with all pools of PtdIns-containing lipids. This is a particular problem in the case of $[^3H]$-inositol labeling experiments for the reasons discussed above. Thus, it was important to determine the $InsP_3$ mass in resting and stimulated platelets. Rittenhouse and Sasson[83] used capillary gas chromatography to measure the mass of $InsP_3$. $InsP_3$ was isolated on a standard anion exchange column and, after several steps, converted to inositol by treatment with alkaline phosphatase. The resulting inositol was treated with silylating reagents to render it volatile. Resting levels of $InsP_3$ were about 20 pmoles/10^9 cells. Thrombin caused a rapid increase in the mass of $InsP_3$ that was detectable at 5 sec and peaked by 15 sec at 150 pmoles/10^9 cells. Assuming an average intracellular platelet volume of 5×10^{-15} l,[108] the resting level of $InsP_3$ would be about 4 μM, and stimulated cells would contain about 30 μM $InsP_3$. Clearly these numbers are inconsistent with expectations from the experiments measuring $InsP_3$-induced Ca^{2+} release, but reasons for this will become apparent. The major disadvantages of this method are the long and intricate sample preparation and the fact that experiments require large amounts of rather concentrated platelet suspensions.

Improved Measurement of InsP₃ Using [³²P]-PO₄ Labeling

As discussed above, and in light of subsequent studies, the method used by Agranoff *et al.*[2] to measure [^{32}P]-InsP$_3$ formation in platelets appeared inadequate. However, since [^{32}P]-labeling has several advantages over other methods, we decided to modify the Agranoff method. In order to determine which phosphorous-containing metabolites might interfere with the measurement of InsP$_3$, several different metabolites known to be found in platelets[42] were electrophoresed in the oxalate buffer system.[25] Only one of the tested metabolites, 2,3,diphosphoglycerate (2,3-DPG), comigrated with InsP$_3$. When the [^{32}P]-labeled metabolites from stimulated platelets that migrated with an InsP$_3$ standard were eluted and chromatographed on paper in two dimensions, only two radioactive spots were seen at positions corresponding to InsP$_3$ and 2,3-DPG standards. The radioactive material at the 2,3-DPG position was totally removed by treatment of platelet extracts with 2,3,-DPG phosphatase. This result implied that if platelet extracts were pretreated with the enzyme, the [^{32}P]-radioactivity at the InsP$_3$ position on electrophoretograms in oxalate buffer would only reflect InsP$_3$. Using this method we found that the resting level of InsP$_3$ was much lower than previously determined by Agranoff *et al.*[2] and that stimulation with thrombin produced at least a tenfold increase in the amount of radioactive InsP$_3$. In addition, we determined that the specific radioactivity of PtdIns-4,5-P$_2$ was the same as that of ATP, indicating that phosphates of these two metabolites were in isotopic equilibrium. This observation was later confirmed in much greater detail by Dr. Holmsen and his colleagues.[99,100] Knowing this and measuring the specific radioactivity of [^{32}P]-ATP, we determined the mass of InsP$_3$ from the measurement of [^{32}P]-InsP$_3$. These measurements were in close agreement with the measurements of Rittenhouse and Sasson.[83] About 23 (+9) pmoles/10^9 cells of InsP$_3$ were found in resting platelets, and 230 (+56) pmoles/10^9 cells of InsP$_3$ were found in thrombin-stimulated platelets.

Inositol 1:2-cyclic, 4,5-trisphosphate

Dawson *et al.*[30] showed that cleavage of PtdIns by PLC could result in either inositol 1-phosphate or inositol 1:2-cyclic phosphate. Cyclic inositol phosphates are acid labile, and inositol 1:2-cyclic phosphate would be converted into the inositol 1-phosphate in the acid extraction systems normally used. Wilson *et al.*[109] showed that ram seminal vesicle PLC could cleave PtdIns 4-P into either Ins(1,4)P$_2$ or inositol 1:2-cyclic 4-bisphosphate and that it could cleave PtdIns 4,5-P$_2$ into either Ins 1,4,5-P$_3$ or inositol 1:2-cyclic 4,5-trisphosphate (Ins 1:2-c4,5-P$_3$). Both of the inositol trisphosphates were active in releasing Ca^{2+} from intact limulus photoreceptors and from saponin-treated platelets.[110] In the case of the limulus photoreceptors, the cyclic InsP$_3$ was five times more potent than its noncyclic counterpart. In the platelet system, Ins 1,4,5-P$_3$ was about 3-5 times more potent than the cyclic form (EC$_{50}$=1.3 μM). They also demonstrated that Ins 1:2-c4,5-P$_3$ would cause release of serotonin from saponin-permeabilized platelets.[111] Connolly *et al.*[21] showed that cyclic inositol phosphates were metabolized in separate pathways (Figure 2). Ins 1:2-c4,5-P_3 is converted only to inositol 1:2-cyclic 4-bisphosphate which is in turn converted only to inositol 1:2-cyclic phosphate which can then be converted to inositol 1-phosphate. Ins 1:2-c4,5-P_3 is a poorer substrate for the 5-phosphatase than is Ins 1,4,5-P_3. Ishii *et al.*[52] demonstrated that platelets could produce Ins 1:2-c4,5-P_3 when stimulated by thrombin. The kinase which phosphorylates Ins 1,4,5-P_3 at the 3 position (see below) does not phosphorylate Ins 1:2-c4,5-P_3.[19] Using this fact, Tarver *et al.*[96] were able to measure the relative masses of the two inositol phosphates formed upon stimulation with thrombin. Ins 1:2-c4,5-P_3 is formed both to a lesser extent and more slowly than Ins 1,4,5-P_3. Thus, these authors concluded that at least during the early stages of cell activation, Ins 1,4,5-P_3 is probably more important as a second messenger than its cyclic counterpart. A similar conclusion has been reached by Dixon and Hokin[33] in the case of pancreatic minilobules. Wong *et al.*[112] failed to find Ins 1:2-c4,5-P_3 in stimulated WRK1 cells and Brown *et al.*[15] were also unsuccessful in their studies on squid retina.

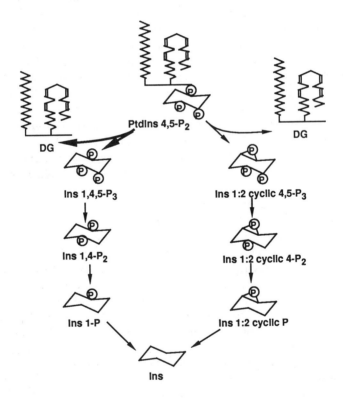

FIGURE 2. Formation and metabolism of inositol 1:2,4,5 cyclic trisphosphate.

ADP-induced InsP₃ Formation

As mentioned above, DP-induced $InsP_3$ release remains somewhat controversial. As a platelet agonist ADP, in contrast to epinephrine, clearly mobilized intracellular Ca^{2+} in the presence or absence of extracellular Ca^{2+}.[38] Vickers *et al.*[101] demonstrated in rabbit platelets that ADP caused a small but significant change in the mass of PtdIns 4,5-P_2 10 sec after agonist addition. A decrease in [^{32}P]-PtdIns 4,5-P_2 was detectable at 60 sec. These experiments suggested that ADP-induced PtdIns 4,5-P_2 metabolism was similar to that induced by thrombin and that ADP-induced intracellular Ca^{2+} mobilization might be due to Ins 1,4,5-P_3. However, Fisher *et al.*[35] failed to detect either a decrease in [^{32}P]-PtdIns 4,5-P_2 or, in a separate experiment, an increase in [^{32}P]-inositol phosphates using [^{32}P]-PO_4-labeled or [^{32}P]-inositol-labeled human platelets, respectively. The same authors showed that ADP increased intracellular Ca^{2+} in quin2-loaded platelets. However, they used different cells for Ca^{2+} measurements than those used for inositol phosphate measurements. They suggested that increases in intracellular Ca^{2+} can occur by a mechanism independent of Ins 1,4,5-P_3-dependent liberation of stored Ca^{2+}.

Our group decided to investigate this question using [^{32}P]-labeled platelets and the modified electrophoretic method described above.[26] An important consideration in our experiments was that the same platelets used to measure $InsP_3$ formation also be used to measure Ca^{2+} mobilization. The platelets studied were preincubated with aspirin to eliminate contributions from cyclooxygenase products. When treated with 20 µM ADP, intracellular Ca^{2+} (measured with fura-2), myosin phosphorylation and $InsP_3$ all increased and reached maximal levels at 5 sec. The increase in $InsP_3$ was 1.7-fold above control and was statistically significant at all times measured. While the ADP-stimulated levels of intracellular Ca^{2+} and myosin phosphorylation were increased in the presence of extracellular Ca^{2+}, there was no statistically significant increase in $InsP_3$ formation when extracellular Ca^{2+} was present. In addition to $InsP_3$, we measured PtdIns 4,5-P_2 and PA formation after ADP stimulation. No significant change in PtdIns 4,5-P_2 was found; however, PA increased twofold. Ionomycin had no effect on PA or $InsP_3$ formation as expected for platelets with an inactive cyclooxygenase.[82] We suggested these results did not support the theory that the mechanism of intracellular Ca^{2+} mobilization for ADP was different than for other platelet agonists. However, we do not exclude the idea that a portion of ADP-stimulated Ca^{2+} mobilization might differ from that of other agonists, possibly by direct opening of Ca^{2+} channels in the plasma membrane as suggested by Sage and Rink.[86]

In a follow-up of their earlier study,[101] Vickers *et al.*[102] measured inositol phosphate metabolism in [^3H]-inositol-labeled ADP-stimulated rabbit platelets. While they found significant increases in InsP and $InsP_2$, a small increase in $InsP_3$ was not statistically significant. These investigators interpreted these data to suggest that PLC was not activated by ADP. However, their explanation seems contrived since the other two inositol phosphates were formed. It is possible that small but unmeasurable amounts of $InsP_3$ were formed and then rapidly converted to other metabolites.

THE METABOLISM OF INS 1,4,5-P_3

Formation of Isomers and Other Polyphosphates

The Formation of Ins 1,4,5-P_3

Most of the early work on inositol phosphate metabolism used rather low-resolution ion exchange chromatography or paper electrophoresis methods to separate the different inositol phosphates. When extracts of [³H]-inositol-labeled rat parotid glands were separated by HPLC ion exchange chromatography, an unexpected result was obtained.[49] In extracts of cells that had been stimulated with carbachol for 15 min, an isomer of the Ins 1,4,5-P_3 was found that was tentatively identified as inositol 1,3,4-trisphosphate (Ins 1,3,4-P_3). The Ins 1,3,4-P_3 isomer was formed after the Ins 1,4,5-P_3 isomer and when the stimulus was removed, the level of Ins 1,3,4-P_3 declined more slowly than that of Ins 1,4,5-P_3.[17,46] These data suggested that Ins 1,3,4-P_3 might be a product formed as a result of Ins 1,4,5-P_3 metabolism; however, at first it was not clear what mechanism would transform one isomer into the other. The source of Ins 1,3,4-P_3 was not phosphatidylinositol 3,4-bisphosphate, since attempts to find this phospholipid proved unsuccessful.[34] Batty *et al.*[7] demonstrated that carbachol-stimulated brain slices contain another new inositol polyphosphate, an inositol tetrakisphosphate, probably inositol 1,3,4,5-tetrakisphosphate (Ins 1,3,4,5-P_4). (The structures of Ins 1,3,4-P_3 and Ins 1,3,4,5-P_4 have been confirmed by Lindon *et al.*[63]) A red cell phosphatase could convert Ins 1,3,4,5-P_4 to Ins 1,3,4-P_3, suggesting that Ins 1,3,4,5-P_4 was the precursor for Ins 1,3,4-P_3. The demonstration of an Ins 1,4,5-P_3 3-kinase activity in rat brain, pancreas and liver[47] indicated that Ins 1,3,4-P_3 could be formed by phosphorylation of Ins 1,4,5-P_3 to yield Ins 1,3,4,5-P_4 followed by phosphatase-catalyzed conversion of the Ins 1,3,4,5-P_4 to Ins 1,3,4-P_3 (Figure 3).

Our studies were the first to demonstrate the existence of this pathway in human platelets.[27] Our approach was to prelabel platelets with [³²P]-PO_4 and the fluorescent Ca^{2+} probe, fura-2, for 1 hr in plasma. Platelets were treated with thrombin (5U/ml) for various lengths of time in a fluorometer in order to continuously monitor changes in intracellular Ca^{2+} and, then they were extracted with perchloric acid. In order to remove interfering nucleotides, extracts were treated with charcoal while pyrophosphate was removed with the enzyme pyrophosphatase. When the extracts were analyzed by HPLC ion exchange chromatography, we found a [³²P]-labeled metabolite comigrating with a [³²P]-labeled Ins 1,4,5-P_3 and one eluting slightly before the Ins 1,4,5-P_3 peak. Also, a [³²P]-labeled peak was found eluting much later than the other two, consistent with an inositol tetrakisphosphate ($InsP_4$). Addition of [³²P]-labeled Ins 1,4,5-P_3 to a lysed platelet preparation demonstrated that platelets contained enzymes

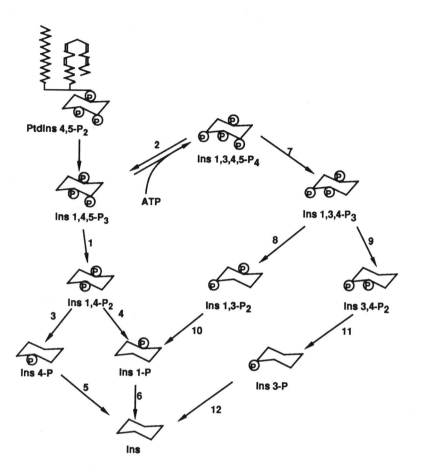

FIGURE 3. The metabolism of inositol 1,4,5-trisphosphate.

to convert Ins 1,3,4-P$_3$ first to an InsP$_4$ and then to an InsP$_3$ isomer that eluded just before Ins 1,4,5-P$_3$. This pattern was exactly that seen in the studies in other cells and thus we identified the InsP$_4$ as Ins 1,3,4,5-P$_4$ and the InsP$_3$ isomer as Ins 1,3,4-P$_3$. Ins 1,4,5-P$_3$ was formed rapidly with detectable increases seen at 1 sec with maximal amounts obtained at 4 sec after the addition of thrombin. It was also rapidly degraded, with close to basal levels being reached after 60 sec. Formation of Ins 1,3,4,5-P$_4$ was almost as rapid as that of Ins 1,4,5-P$_3$; however, Ins 1,3,4,5-P$_4$ was more slowly removed, with detectable levels remaining after 300 sec. Ins 1,3,4-P$_3$ was formed after an initial lag of 2 sec and reached its maximal level at 30 sec. The amount of [^{32}P]-labeled Ins 1,3,4-P$_3$ formed was tenfold higher than that of the 1,4,5 isomer. Based on the previously measured mass of total InsP$_3$ formed under similar conditions,[25,83] it was calculated that Ins 1,4,5-P$_3$ concentration reached about 2 μM with thrombin stimulation. Comparing the thrombin-stimulated changes in intracellular Ca^{2+}

concentration with the time course of inositol polyphosphate metabolism indicated that elevated Ca^{2+} concentrations were maintained for longer periods of time than could be explained by the concentration of Ins 1,4,5-P_3 alone. Therefore, we suggested that either Ins 1,3,4,5-P_4 or Ins 1,3,4-P_3 might be responsible for the sustained maintenance of elevated intracellular Ca^{2+} concentrations in thrombin-stimulated platelets.

Tarver *et al.*[96] essentially confirmed these findings. They devised an assay for the mass of Ins 1,4,5-P_3 that depended on the ability of Ins 1,4,5-P_3 3-kinase to convert Ins 1,4,5-P_3 to radioactive Ins 1,3,4,5-P_4 using [^{32}P]-labeled ATP. The difference between the total InsP$_3$ and Ins 1,4,5-P_3 was used to calculate Ins 1,3,4-P_3 formation. The amount of Ins 1,4,5-P_3 formed was 10-20 pmol/10^9 cells (2-4 µM), and the amount of Ins 1,4,5-P_3 was tenfold higher. Culty *et al.*[24] also demonstrated the presence of Ins 1,3,4-P_3 in thrombin-stimulated platelets labeled with [^{32}P]-inositol.

The Regulation of Inositol 1,3,4-trisphosphate 3-kinase by Calcium

In 1986, Lew *et al.*[62] showed that fMet-Leu-Phe-stimulated HL-60 cells produced the same amount of Ins 1,4,5-P_3 but less Ins 1,3,4-P_3 if the intracellular Ca^{2+} stores of the cells had been previously depleted. This experiment suggested that Ca^{2+} might play a role in regulating the metabolic conversion of Ins 1,4,5-P_3 to Ins 1,3,4-P_3. Rossier *et al.*[85] used permeabilized preparations of both adrenal glomerulosa cells and rat aortic smooth-muscle cells,[84] and Biden and Wollheim[11] used broken preparations of insulin-secreting Finm5F cells to show that the conversion of Ins 1,4,5-P_3 to Ins 1,3,4-P_3 is augmented by increasing Ca^{2+} over the physiological range. Biden and Wollheim[11] demonstrated that the 3-kinase was Ca^{2+}-dependent with an EC_{50} of 0.8 µM. Later, partially purified kinase preparations were obtained from a malignant human T-cell line[44] and pig aortic smooth muscle. Both enzymes were found to be regulated by Ca^{2+}. The purest preparation of the smooth muscle kinase was no longer Ca^{2+}-dependent, but Ca^{2+} sensitivity was restored by the addition of calmodulin. A similar kinase has been found in the particulate fraction of lysed turkey erythrocytes.[76] Johanson *et al.*[54] reported a 4000-fold purification of rat brain Ins(1,4,5)P_3 3-kinase. This enzyme was initially activated by Ca^{2+} but lost Ca^{2+} sensitivity during purification. Calmodulin restored Ca^{2+} sensitivity to the kinase. This is in contrast to studies by Irvine *et al.*[47] who failed to find Ca^{2+} regulation of the brain enzyme.

In the case of human platelets, we observed that more total InsP$_3$ was formed when these cells were stimulated with agonists (9,11 azo-PGH2 or thrombin) in the presence of extracellular Ca^{2+} than in the absence of extracellular Ca^{2+}.[28] HPLC analysis of the InsP$_3$ indicated that the extra InsP$_3$ formed in the presence of extracellular Ca^{2+} was primarily Ins 1,3,4-P_3. Additional Ins 1,4,5-P_3 was also formed in the presence of extracellular Ca^{2+}. Using electrically permeabilized platelets, we showed that conversion of [^{3}H]-Ins(1,4,5)P$_3$ to [^{3}H]InsP$_4$ in platelets was Ca^{2+} dependent with an EC$_{50}$ at 2.5 µM Ca^{2+}. By contrast, dephosphorylation of [^{3}H]InsP$_4$ to [^{3}H]Ins(1,3,4)P$_3$ was not activated by Ca^{2+}. A partially purified preparation of Ins(1,4,5)P$_3$ 3-kinase from human platelets was found to be insensitive to Ca^{2+}, but addition of calmodulin restored Ca^{2+} sensitivity to the kinase, increasing its activity about fivefold. These results show that in human platelets, as in other cells, the metabolism of Ins(1,4,5)P$_3$ is regulated by Ca^{2+}-calmodulin.

The Function of Ins 1,3,4,5-P$_4$

At present, the function of neither Ins 1,3,4-P$_3$ nor Ins 1,3,4,5-P$_4$ is known. Both the fact that ATP is required in this pathway and the fact that the 3-kinase is regulated by calmodulin suggest that these metabolites must be important. Irvine and his collaborators[50,76] have proposed that Ins 1,3,4,5-P$_4$ allows Ca^{2+} entry at special gates in the plasma membrane. This effect requires the presence of InsP$_3$, and in these studies Ins 2,4,5-P$_3$ was microinjected into the egg cells since its Ca^{2+}-mobilizing activity is similar to that of Ins 1,4,5-P$_3$ but is poorly converted by the 3-kinase to Ins 1,3,4,5-P$_4$. A similar synergism has been seen in mouse lacrimal glands[76] and Aplysia neurons[51] (Kaczmarek, unpublished data). However, Crossley et al.[22] failed to detect this synergism and concluded that extracellular Ca^{2+} is not important in sea urchin egg activation. Irvine et al.[51] have proposed that Ins 1,3,4,5-P$_4$, in the presence of GTP, controls the entry of Ca^{2+} from the extracellular fluid and from some pools of endoplasmic reticulum into the special pool of endoplasmic reticulum (calciosome) that is gated by Ins 1,4,5-P$_3$. For a full discussion of the evidence that leads to this conclusion, see the review of Irvine et al.[51] There is at present no evidence for such a mechanism in platelets.

Some evidence that Ins 1,4,5-P$_3$ itself can open Ca^{2+} channels in the plasma membrane has been put forth. Using patch clamp techniques, Kuno and Gardner[59] have shown that Ins 1,4,5-P$_3$ opens voltage-insensitive channels in human T-lymphocytes. Plasma membrane-derived vesicles from platelets released ^{45}Ca^{2+} when treated with Ins 1,4,5-P$_3$.[81] This release was dependent on the Ins 1,4,5-P$_3$ concentration but was not saturated with 20 µM Ins 1,4,5-P$_3$. In light of this requirement for a high concentration of Ins 1,4,5-P$_3$, it is hard to know if this mechanism is of physiological importance.

Another proposal for the action of Ins 1,3,4,5-P$_4$ is that it potentiates the action of Ins 1,4,5-P$_3$ by increasing the duration of the Ins 1,4,5-P$_3$-induced Ca^{2+} transient.[55] However, high concentrations of Ins 1,3,4,5-P$_4$, probably outside the range formed in cells, were needed to obtain a substantial effect. Ins 1,3,4,5-P$_4$ receptor binding has been demonstrated in different rat tissue membranes.[97] [^{3}H]-labeled Ins 1,3,4,5-P$_4$ was displaced by both Ins 1,4,5-P$_3$ and Ins 1,3,4,5-P$_4$. The relative affinity varied with the tissue, and in most cases, Ins 1,3,4,5-P$_4$ was more effective in displacing the radiolabel.

A Possible Function of Ins 1,3,4-P$_3$

As noted above, much greater quantities of Ins 1,3,4-P$_3$ are formed in thrombin-stimulated platelets than Ins 1,4,5-P$_3$.[27,96] Irvine *et al.*[48] found that Ins 1,3,4-P$_3$ could release stored Ca^{2+} from intracellular stores but was 30 times less potent with an EC_{50} of 9 μM.[48] Since thrombin-stimulated levels of Ins 1,3,4-P$_3$ reached levels greater than 9 μM (20-40 μM), we decided to determine the EC_{50} for Ca^{2+} release in saponin-permeabilized human platelets. Platelets were treated with saponin in the presence of an ATP-regenerating system and the Ca^{2+} indicator quin2. Addition of the saponin resulted in a decrease in measured Ca^{2+}, presumably resulting from action of intracellular pumps. Addition of either Ins 1,4,5-P$_3$ or Ins 1,3,4-P$_3$ caused a dose-dependent increase in the measured Ca^{2+} concentration. The results of several such experiments are summarized in Figure 4. The EC_{50} for Ca^{2+} release by Ins 1,4,5-P$_3$ was 0.125 μM, and that for Ins 1,3,4-P$_3$ was 20 μM, a concentration that can be attained in thrombin-stimulated platelets. These data suggest that Ins 1,3,4-P$_3$ may prolong the Ca^{2+} transient in platelets under some circumstances. However, one should consider the results as preliminary until we can eliminate the possibility that a small amount of Ins 1,4,5-P$_3$ was contaminating our Ins 1,3,4-P$_3$ preparation.

The Pathways in the Formation of Inositol from Ins 1,4,5-P$_3$

Figure 3 shows the pathways that lead to the degradation of Ins 1,4,5-P$_3$ and Ins 1,3,4-P$_3$ to inositol. The evidence for these pathways is fully discussed in the review by Irvine *et al.*[51] and thus will not be discussed in detail here. While this scheme looks complicated, only four major phosphatases seem to be involved. First, the 5 phosphate is removed by a 5-phosphatase (reactions 1 and 7 of Figure 3). The same enzyme removes the 1 phosphate from Ins 1,4-P$_2$ as removes it from Ins 1,3,4-P$_3$ (reactions 3 and 9), and this enzyme is inhibited by lithium.[45] Note that this enzyme does not use Ins 1,4,5-P$_3$ as a substrate.[45] Alternatively, the 4-phosphate can be removed first (reactions 4 and 8), but it is not clear whether this is catalyzed by a single enzyme.[51] Ins 1,3,4-P$_3$ is also degraded by removal of the 4 phosphate (reaction 11). A single inositol monophosphatase may remove the final phosphate from all inositol monophosphates (reactions 5, 6 and 12).[51]

Other Inositol Polyphosphates

While this picture of the metabolism of the inositol phosphates is quite complex, recent evidence suggests that additional complexities exist in cells. A kinase activity has been demonstrated in rat liver[87] and bovine adrenal glomerulosa cells[5] that phosphorylates Ins 1,3,4-P$_3$ to Ins 1,3,4,6-P$_4$. Stephens *et al.*[91] showed that another inositol tetrakisphosphate, Ins 1,4,5,6-P$_4$, was found in both murine macrophages and chick erythrocytes. The level of Ins 1,4,5,6-P$_4$ was not altered by stimulation of the macrophages with PAF; however, more Ins 1,3,4,5-P$_4$ was formed. A kinase has been partially purified from murine

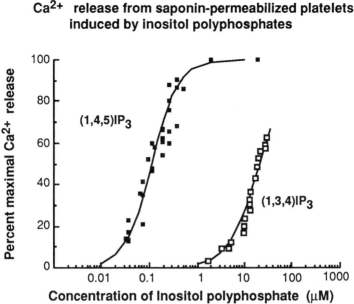

FIGURE 4. Dose-response of release of intracellular Ca^{2+} from saponin-treated platelets. Washed human platelets (3×10^8 cells/ml) were permeabilized with 30 mg/ml of saponin. An ATP regenerating system consisting of 10 μM creatine phosphate, 10 U/ml of creating phosphokinase and 10 μM MgATP was present. Free Ca^{2+} was monitored by the fluorescence of quin2 free acid (10 μM). All samples were individually calibrated and response measured as the change in the Ca^{2+} concentration produced by a given dose of inositol phosphate.

macrophages than can phosphorylate Ins 1,4,5,6-P_4 to form an inositol pentakisphosphate, Ins 1,3,4,5,6-P_5.[92] $InsP_5$ and $InsP_6$ have been found in mammalian brain tissue.[39] It has been suggested that these inositol polyphosphates may be in compartments separate from the cytosol and might function as neurotransmitters. Recently, several different isomers of $InsP_4$ and Ins 1,3,4,5,6-P_5 were found in thrombin-stimulated human platelets.[71]

The Effect of Activation of Protein Kinase C on Inositol Phosphate Metabolism

The first indication that protein kinase C might have an effect on PtdIns metabolism came from experiments that showed that addition of either the synthetic diacylglycerol, 1-oleoyl-2-acetate-glycerol (OAG), or phorbol myristate acetate (PMA) would stimulate an increase in [^{32}P]-PtdIns 4,5-P_2 and [^{32}P]-PtdIns 4-P when they were added to [^{32}P]-labeled platelets.[31,36] In the latter study, this change in radiolabeled metabolites was shown to be accompanied by a similar increase in mass of both PtdIns 4,5-P_2 and PtdIns 4-P. A fall in both mass and [^{32}P]-labeled PtdIns was also seen. Watson and Lapetina[103] showed that both OAG and PMA depressed the amount of inositol phosphates formed when [^{32}P]-inositol-labeled platelets were stimulated by thrombin. This observation was confirmed by Rittenhouse and Sasson[83] who showed that the mass of InsP$_3$ formed in thrombin-stimulated platelets was reduced by the addition of PMA. A protein kinase C inhibitor, H-7, was shown to inhibit the effects of PMA on thrombin-induced PtdIns metabolism.[98] MacIntyre *et al.*[66] showed that PMA inhibited the agonist-induced increase in intracellular Ca^{2+} and PA formation. Poll *et al.*[79] demonstrated that PMA inhibited the rise of intracellular Ca^{2+} induced by NaF. These results were confirmed by Yoshida *et al.*[113] who added PMA to platelets 10 sec prior to the addition of ADP or thrombin. A dose-dependent inhibition of shape change, association of myosin with cytoskeletons and the rise of intracellular Ca^{2+} concentrations were observed. Adding PMA after the agonist increased the rate of Ca^{2+} fall to basal levels.

Different mechanisms have been proposed to explain the inhibitory effects of PMA on agonist-stimulated PtdIns turnover and Ca^{2+} mobilization. Connolly *et al.*[20] suggested that inositol trisphosphate 5'-phosphomonoesterase was the major protein kinase C substrate, known as either P40 or P47. This enzyme was found to be activated when it was phosphorylated by protein kinase C. In support of this hypothesis, Molina y Vedia and Lapetina[75] found that PMA increased the rate of degradation of InsP$_3$. However, the identity of P40 as inositol trisphosphate 5'-phosphomonoesterase has been questioned.[40] Furthermore, Kennedy *et al.*[57] have failed to confirm the fact that rabbit neutrophil inositol trisphosphate 5'-phosphomonoesterase is activated by protein kinase C phosphorylation. As an alternative, it was shown that protein kinase C could phosphorylate a 41 kDa platelet protein (presumably the a subunit of G$_i$) that is also a substrate for ADP ribosylation by pertussis toxin.[37] Crouch and Lapetina[23] have suggested that G$_i$ plays a role in the coupling of the thrombin receptor to activation of PLC, and that phosphorylation of α_i by protein kinase C would block thrombin-induced activation of PLC. Furthermore, they suggest that this effect explains the desensitization that occurs when platelets are stimulated a second time by thrombin. However, Banga *et al.*[6] have concluded that while ribosylation of a 41 kDa platelet protein interfered with the ability of thrombin to inhibit adenylate cyclase, it had no effect on the activation of PLC by thrombin. Watson *et al.*[105] used the protein kinase inhibitor staurosporine to study the effects of protein phosphorylation on platelet function and found that while 1 µM staurosporine blocked thrombin-induced protein phosphorylation, it had no effect on thrombin-induced PA formation. These authors suggest that agents such as phorbol esters may induce phosphorylations not normally seen in cells activated by more physiologic agonists.

CONCLUDING REMARKS

Significant progress has been made in the last five years in defining the role of the "PI response" and inositol polyphosphates in cell activation. The theory that Ins 1,4,5-P$_3$ is an important second messenger and that it causes release of

Ca^{2+} from internal stores has been supported by a great deal of evidence. However, several key aspects of this theory are missing. For example, the receptor for Ins 1,4,5-P_3 in internal membranes has not been fully defined. How does Ins 1,4,5-P_3 interact with this receptor, and what is the nature of the Ca^{2+} channel that is opened to allow the escape of stored Ca^{2+}? Another important question is: What is the nature of the channels that allow entry of extracellular Ca^{2+}? Putney[80] has put forth a theory in which depletion of intracellular Ca^{2+} stores following Ins 1,4,5-P_3-induced release allows entry of extracellular Ca^{2+}; however, there is little direct proof for this model.

The metabolism of Ins 1,4,5-P_3 is extremely complicated. It remains to be determined whether there is a role for cyclic inositol phosphates in cell function. Energy in the form of ATP is used to make inositol tetrakis- and pentakisphosphates, and the 3-kinase is regulated by calmodulin. These observations suggest that these pathways are of importance to the function of the cell but exactly how has not been determined. Platelets stimulated with thrombin produce a large amount of Ins 1,3,4-P_3. Does this metabolite have any role in platelet function? We have proposed that this isomer may have a modulatory effect on the cytosolic free Ca^{2+} concentration sustaining high Ca^{2+} concentrations for prolonged periods of time after agonist stimulation.

Even less clear is the effect, if any, of the protein kinase C system on inositol phosphate production. Is phosphorylation of a G-unit responsible for receptor desensitization?

We have finished with more questions than answers; however, based on the rapid progress that has been made in this field in the last few years, we do not expect to wait long for the answers.

List of Abbreviations

OAG = 1-oleoyl-2-acetate-glycerol
PA = Phosphatidic acid
PAF = Platelet-activating factor
PMA = Phorbol myristate acetate
PtdIns = Phosphatidylinositol
PtdIns 4-P = Phosphatidylinositol 4-phosphate
PtdIns 4,5-P_2 = Phosphatidylinositol 4,5-bisphosphate
Ins(1,4)P_2 = Inositol 1,4-bisphosphate
PLC = Phospholipase C
InsP$_3$ = Inositol trisphosphate (isomeric form unspecified)
Ins 1,4,5-P_3 = Inositol 1,4,5-trisphosphate

(Note: the abbreviations of other inositol phosphates are based on this format with changes to reflect the positions and numbers of phosphates).

ACKNOWLEDGMENTS

This work was supported by grants HL-14217 and BSRG S07 RR05417 from the NIH. We thank Carol Dangelmaier, Sandra Gordon and Bryan Smith for collaboration on the work included in this manuscript.

REFERENCES

1. **Adunyah, S.E., and Dean, W.L.,** Inositol triphosphate-induced Ca^{2+} release from human platelet membranes, *Biochem. Biophys. Res. Commun.*, 128, 1274-1280, 1985.
2. **Agranoff, B.W., Murphy, P., and Sequin, E.B.,** Thrombin-induced phosphodiesterase cleavage of phosphatidylinositol bisphosphate in human platelets, *J. Biol. Chem.*, 258, 2076-2078, 1983.
3. **Authi, K.S., and Crawford, N.,** Inositol 1,4,5-trisphosphate-induced release of sequestered Ca^{2+} from highly purified human platelet intracellular membranes, *Biochem. J.*, 230, 247-253, 1985.
4. **Authi, K.A., Hornby, E.J., Evenden, B.J., and Crawford, B.J.,** Inositol 1,4,5-trisphosphate (IP_3) induced rapid formation of thromboxane B2 in saponin-permeabilized human platelets: Mechanism of IP_3 action, *FEBS Lett.*, 213, 95-101, 1987.
5. **Balla, T., Guillemette, G., Baukal, A.J., and Catt, K.J.,** Formation of inositol 1,3,4,6-tetrakisphosphate during angiotensin II action in bovine adrenal glomerulosa cells, *Biochem. Biophys. Res. Commun.*, 148, 199-205, 1987.
6. **Banga, H.S., Walker, R.K., Winberry, L.K., and Rittenhouse, S.E.,** Platelet adenylate cyclase and phospholipase C are affected differentially by ADP-ribosylation. Effects on thrombin-mediated responses, *Biochem. J.*, 252, 297-300, 1988.
7. **Batty, I.R., Nahorski, S.R., and Irvine, R.F.,** Rapid formation of inositol 1,3,4,5-tetrakisphosphate following muscarinic receptor stimulation of rat cerebral cortical slices, *Biochem. J.*, 232, 211-215, 1985.
8. **Berridge, M.J., Dawson, R.M., Downes, C.P., Heslop, J.P., and Irvine, R.F.,** Changes in the levels of inositol phosphates after agonist-dependent hydrolysis of membrane phosphoinositides, *Biochem. J.*, 212, 473-482, 1983.
9. **Berridge, M.L.,** Inositol trisphosphate and diacylglycerol as second messengers, *Biochem. J.*, 220, 345-360, 1984.
10. **Berridge, M.L., and Irvine, R.F.,** Inositol trisphosphate, a novel second messenger in cellular signal transduction, *Nature*, 312, 315-321, 1984.
11. **Biden, T.J., and Wollheim, C.B.,** Ca^{2+} regulates the inositol tris/tetrakisphosphate pathway in intact and broken preparations of insulin-secreting RINm5F cells, *J. Biol. Chem.*, 261, 11931-11934, 1986.
12. **Billah, M.M., and Lapetina, E.G.,** Rapid decrease of phosphatidylinositol 4,5 bisphosphate in thrombin-stimulated platelets, *J. Biol. Chem.*, 257, 12705-12708, 1982.
13. **Brass, L.F., and Joseph, S.K.,** A role for inositol triphosphate in intracellular Ca^{2+} mobilization and granule secretion in platelets, *J. Biol. Chem.*, 260, 15172-15179, 1985.
14. **Brass, L.F., Shaller, C.C., and Belmonte, E.J.,** Inositol 1,4,5-triphosphate-induced granule secretion in platelets: Evidence that the activation of phospholipase C mediated by platelet thromboxane receptors involves a guanine nucleotide binding protein-dependent mechanism distinct from that of thrombin, *J. Clin. Invest.*, 79, 1269-1275, 1987.
15. **Brown, J.E., Rudnick, M., Letcher, A.J., and Irvine, R.F.,** Formation of methylphosphoryl inositol phosphates by extractions that employ methanol, *Biochem. J.*, 253, 703-710, 1988.
16. **Burgess, G.M., Godfrey, P.P., McKinney, J.S., Berridge, M.J., Irvine, R.F., and Putney, J.W., Jr.,** The second messenger linking receptor activation to internal Ca release in liver, *Nature*, 309, 63-66, 1984.

17. **Burgess, G.M., McKinney, J.S., Irvine, R.F., and Putney, J.W., Jr.,** Inositol 1,4,5-trisphosphate and inositol 1,3,4-trisphosphate formation in Ca^{2+}-mobilizing-hormone-activated cells, *Biochem. J.*, 232, 237-243, 1985.

18. **Cohen, P., and Derksen, A.,** Comparison of phospholipid and fatty acid composition of human erythrocytes and platelets, *Br. J. Haematol.*, 17, 359-371, 1969.

19. **Connolly, T.M., Bansal, V.S., Bross, T.E., Irvine, R.F., and Majerus, P.W.,** The metabolism of tris- and tetraphosphates of inositol by 5-phosphomonoesterase and 3-kinase enzymes, *J. Biol. Chem.*, 262, 2146-2149, 1987.

20. **Connolly, T.M., Lawing, W.J., Jr., and Majerus, P.W.,** Protein kinase C phosphorylates human platelet inositol trisphosphate 5' monoesterase, increasing the phosphatase activity, *Cell*, 46, 951-958, 1986.

21. **Connolly, T.M., Wilson, D.B., Bross, T.E., and Majerus, P.W.,** Isolation and characterization of the inositol cyclic phosphate products of phosphoinositide cleavage by phospholipase C. Metabolism in cell-free extracts, *J. Biol. Chem.*, 261, 122-126, 1986.

22. **Crossley, I., Swann, K., Chambers, E., and Whitaker, M.,** Activation of sea urchin eggs by inositol phosphates is independent of external calcium, *Biochem. J.*, 252, 257-262, 1988.

23. **Crouch, M.F., and Lapetina, E.G.,** A role for G_i in control of thrombin receptor-phospholipase C coupling in human platelets, *J. Biol. Chem.*, 263, 3363-3371, 1988.

24. **Culty, M., Davidson, M.M., and Haslam, R.J.,** Effects of guanosine 5'-[gamma-thio]triphos phate and thrombin on the phosphoinositide metabolism of electropermeabilized human platelets, *Eur. J. Biochem.*, 171, 523-533, 1988.

24. **Dangelmaier, C.A., Daniel, J.L., and Smith, J.B.,** Determination of basal and stimulated levels of inositol triphosphate in [^{32}P]orthophosphate-labeled platelets, *Anal. Biochem.*, 154, 414-419, 1986.

26. **Daniel, J.L., Dangelmaier, C.A., Selak, M., and Smith, J.B.,** ADP stimulated IP_3 formation in human platelets, *FEBS Lett.*, 206, 299-303, 1986.

27. **Daniel, J.L., Dangelmaier, C.A., and Smith, J.B.,** Formation and metabolism of inositol 1,4,5-trisphosphate in human platelets, *Biochem. J.*, 246, 109-114, 1987.

28. **Daniel, J.L., Dangelmaier, C.A., and Smith, J.B.,** Calcium modulates the generation of inositol 1,3,4-trisphosphate in human platelets by the activation of inositol 1,4,5-trisphosphate 3-kinase, *Biochem. J.*, 253, 789-794, 1988.

29. **Daniel, J.L., Molish, I.R., and Holmsen, H.,** Radioactive labeling of the adenine nucleotide pool of cells as a method to distinguish among intracellular compartments. Studies on platelets, *Biochim. Biophys. Acta*, 632, 444-453, 1980.

30. **Dawson, R.M.C., Freinkel, N., Jungawala, F.B., and Clarke, N.,** The enzymatic formation of myoinositol 1:2 cyclic phosphate from phosphatidylinositol, *Biochem. J.*, 122, 605-607, 1971.

31. **de Chaffoy de Courcelles, D., Roevens, P., and Van Belle, H.,** 12-O-Tetradecanoylphorbol 13-acetate stimulates inositol lipid phosphorylation in intact human platelets, *FEBS Lett.*, 173, 389-393, 1984.

32. **de Chaffoy de Courcelles, D., Roevens, P., Wynants, J., and Van Belle, H.,** Serotonin-induced alterations in inositol phospholipid metabolism in human platelets, *Biochim. Biophys. Acta*, 927, 291-302, 1987.

33. **Dixon, J.F., and Hokin, L.E.,** Inositol 1,2-cyclic 4,5-trisphosphate concentration relative to inositol 1,4,5-trisphosphate in pancreatic minilobules on stimulation with carbamylcholine in the absence of lithium: Possible role as a second messenger in long- but not short-term responses, *J. Biol. Chem.*, 262, 13892-13895, 1987.

34. **Downes, C.P., Hawkins, P.T., and Irvine, R.F.,** Inositol 1,3,4,5-tetrakisphosphate and not phosphatidylinositol 3,4-bisphosphate is the probable precursor of inositol 1,3,4-trisphosphate in agonist-stimulated parotid gland, *Biochem. J.*, 238, 501-506, 1986.

35. **Fisher, G.J., Bakshian, S., and Baldassare, J.J.,** Activation of human platelets by ADP causes a rapid rise in cytosolic free calcium without hydrolysis of phosphatidylinositol 4,5-bisphosphate, *Biochem. Biophys. Res. Commun.*, 129, 958-964, 1985.

36. **Halenda, S.P. and Feinstein, M.B.**, Phorbol myristate acetate stimulates formation of phospatidylinositol 4-phosphate and phosphatidylinositol 4,5-bisphosphate in human platelets, *Biochem. Biophys. Res. Commun.*, 124, 517-513, 1984.

37. **Halenda, S.P., Volpi, M., Zavoico, G.B., Sha'afi, R.I., and Feinstein, M.B.**, Effect of thrombin, phorbol myristate acetate and prostaglandin D2 on the 40-41 kDa protein that is ADP ribosylated by pertussis toxin in platelets, *FEBS Lett.*, 204, 342-346, 1986.

38. **Hallam, T.J., and Rink, T.J.**, Responses to adenosine diphosphate in human platelets loaded with the fluorescent calcium indicator quin2, *J. Physiol.* (London), 368, 131-146, 1985.

39. **Hanley, M.R., Jackson, T.R., Vallejo, M., Patterson, S.I., Thastrup, O., Lightman, S., Rodgers, J., Henderson, G., and Pini, A.**, Neural function: Metabolism and action of inositol metabolites in mammalian brain, *Philos. Trans. R. Soc. Lond. [Biol.]*, 320, 381-398, 1988

40. **Haslam, R.J.**, Signal transduction in platelet activation, in *Thrombosis and Haemostasis 1987*, Verstraete, M., Vermylen, J., Lijnen, H.R., and Arnout, J., Eds., International Society on Thrombosis and Haemostasis and Leuven University Press, Leuven, 1987, 147-174.

41. **Hokin, L.E., and Hokin, M.R.**, Effects of acetylcholine on phosphate turnover in phospholipids of brain cortex *in vitro*, *Biochim. Biophys. Acta*, 16, 229-237, 1955.

42. **Holmsen, H., Akkerman, J.W.N., and Dangelmaier, C.A.**, Determination of levels of glycolytic intermediates and nucleotides in platelets by pulse-labeling with [^{32}P]orthophosphate, *Anal. Biochem.*, 131, 266-272, 1983.

43. **Huang, E.M., and Detwiler, T.C.**, Stimulus-response coupling mechanisms, in *Biochemistry of Platelets*, Phillips, D.R., and Shuman, M.A., Eds., Academic Press, New York, 1986, 2-68.

44. **Imboden, J.B., and Pattison, G.**, Regulation of inositol 1,4,5-trisphosphate kinase activity after stimulation of human T cell antigen receptor, *J. Clin. Invest.*, 79, 1538-1541, 1987.

45. **Inhorn, R.C., Bansal, V.S., and Majerus, P.W.**, Pathway for inositol 1,3,4-trisphosphate and 1,4-bisphosphate metabolism, *Proc. Natl. Acad. Sci. USA*, 84, 2170-2174, 1987.

46. **Irvine, R.F., Anggard, E.E., Letcher, A.J., and Downes, C.P.**, Metabolism of inositol 1,4,5-trisphosphate and inositol 1,3,4-trisphosphate in rat parotid glands, *Biochem. J.*, 229, 505-511, 1985.

47. **Irvine, R.F., Letcher, A.J., Heslop, J.P., and Berridge, M.J.**, The inositol tris/tetrakisphosphate pathway—demonstration of Ins (1,4,5)P$_3$ 3-kinase activity in animal tissues, *Nature*, 320, 631-634, 1986.

48. **Irvine, R.F., Letcher, A.J., Lander, D.J., and Berridge, M.J.**, Specificity of inositol phosphate-stimulated Ca^{2+} mobilization from Swiss-mouse 3T3 cells, *Biochem. J.*, 240, 301-304, 1986.

49. **Irvine, R.F., Letcher, A.J., Lander, D.J., and Downes, C.P.**, Inositol trisphosphates in carbachol-stimulated rat parotid glands, *Biochem. J.*, 223, 237-243, 1984.

50. **Irvine, R.F., and Moor, R.M.**, Micro-injection of inositol 1,3,4,5-tetrakisphosphate activates sea urchin eggs by a mechanism dependent on external Ca^{2+}, *Biochem. J.*, 240, 917-920, 1986.

51. **Irvine, R.F., Moor, R.M., Pollock, W.K., Smith, P.M., and Wreggett, K.A.**, Inositol phosphates: Proliferation, metabolism and function, *Philos. Trans. R. Soc. Lond. [Biol.]*, 320, 281-298, 1988.

52. **Ishii, H., Connolly, T.M., Bross, T.E., and Majerus, P.W.**, Inositol cyclic triphosphate [inositol 1,2-(cyclic)-4,5-triphosphate] is formed upon thrombin stimulation of human platelets, *Proc. Natl. Acad. Sci. USA*, 83, 6397-6401, 1986.

53. **Israels, S.J., Robinson, P., Dockerty, J.D., and Gerrard, J.M.**, Activation of permeabilized platelets by inositol-1,4,5-trisphosphate, *Thromb. Res.*, 40, 499-509, 1985.

54. **Johanson, R.A., Hansen, C.A., and Williamson, J.R.**, Purification of D-myo-inositol 1,4,5-trisphosphate 3-kinase from rat brain, *J. Biol. Chem.*, 263, 7465-7471, 1988.

55. **Joseph, S.K., Hansen, C.A., and Williamson, J.R.** Inositol 1,3,4,5-tetrakisphosphate increases the duration of the inositol 1,4,5-trisphosphate-mediated Ca^{2+} transient, *FEBS Lett.*, 219, 125-129, 1987.

56. **Joseph, S.K., Thomas, A.P., Williams, R.J., Irvine, R.F., and Williamson, J.R.,** Myo-inositol 1,4,5-trisphosphate. A second messenger for the hormonal mobilization of intracellular Ca^{2+} in liver, *J. Biol. Chem.*, 259, 3077-3081, 1984.

57. **Kennedy, S.P., Sha'afi, R.I., and Becker, E.L.,** Demonstration of inositol phosphate 5-phosphomonoesterase activity in rabbit neutrophils: Absence of a role for protein kinase C, *Biochem. Biophys. Res. Commun.*, 155, 189-196, 1988.

58. **Kirk, C.J., and Michell, R.H.,** Phosphatidylinositol metabolism in rat hepatocytes stimulated by vasopressin, *Biochem. J.*, 194, 155-165, 1981.

59. **Kuno, M., and Gardner, P.,** Ion channels activated by inositol 1,4,5-trisphosphate in plasma membrane of human T-lymphocytes, *Nature*, 326, 301-304, 1987.

60. **Labarca, R., Janowsky, A., and Paul, S.M.,** Manganese stimulate incorporation [³H]inositol into an agonist-insensitive pool of phosphatidylinositol in brain membranes, *Biochem. Biophys. Res. Commun.*, 132, 540-547, 1985.

61. **Lapetina, E.G., Watson, S.P., and Cuatrecasas, P.,** Myo-inositol 1,4,5-trisphosphate stimulates protein phosphorylation in saponin-permeabilized human platelets, *Proc. Natl. Acad. Sci. USA*, 81, 7431-7435, 1984.

62. **Lew, P.D., Monod, A., Krause, K.H., Waldvogel, F.A., Biden, T.J., and Schlegel, W.,** The role of cytosolic free calcium in the generation of inositol 1,4,5-trisphosphate and inositol 1,3,4-trisphosphate in HL-60 cells: Differential effects of chemotactic peptide receptor stimulation at distinct Ca^{2+} levels, *J. Biol. Chem.*, 261, 13121-13127, 1986.

63. **Lindon, J.C., Baker, D.J., Williams, J.M., and Irvine, R.F.,** Confirmation of the identities of inositol 1,3,4-trisphosphate and inositol 1,3,4,5-tetrakisphosphate by the use of one-dimensional and two-dimensional n.m.r. spectroscopy, *Biochem. J.*, 244, 591-595, 1987.

64. **Lloyd, J.V., and Mustard, J.F.,** Changes in ³²P-content of phosphatidic acid and the phosphoinositides of rabbit platelets during aggregation induced by collagen or thrombin, *Br. J. Haematol.*, 26, 243-253, 1974.

65. **Lloyd, J.V., Nishizawa, E.E., Joist, J.H., and Mustard, J.F.,** The effect of ADP-induced aggregation on ³²PO₄ incorporation into phosphatic acid and phosphoinositides of rabbit platelets, *Br. J. Haematol.*, 25, 589-604, 1973.

66. **MacIntyre, D.E., McNicol, A., and Drummond, A.H.,** Tumour-promoting phorbol esters inhibit agonist-induced phosphatidate formation and Ca^{2+} flux in human platelets, *FEBS Lett.*, 180, 160-164, 1985.

67. **MacIntyre, D.E., and Pollock, W.K.,** Platelet-activating factor stimulates phosphatidylinositol turnover in human platelets, *Biochem. J.*, 212, 433-437, 1983.

68. **MacIntyre, D.E., Pollock, W.K., Shaw, A., Bushfield, M., MacMillan, L.J., and McNichol, A.,** Agonist-induced inositol phospholipid metabolism and Ca⁺⁺ flux in human platelet activation, in *Mechanism of Stimulus-Response Coupling in Platelets*, Westick, J., Scully, M.F., MacIntyre, D.E., and Kakkar, V.V., Eds., Plenum Press, New York and London, 1985, 127-144.

69. **Marcus, A.J., Ullman, H.J., and Safier, L.B.,** Lipid composition of subcellular particles of human blood platelets, *J. Lipid Res.*, 10, 108-114, 1969.

70. **Mauco, G., Chap, H., and Douste-Blazy, L.,** Platelet-activating factor promotes an early degradation of phosphatidylinositol-4,5-bisphosphate in rabbit platelets, *FEBS Lett.*, 153, 361-365, 1983.

71. **Mayr, G.W.,** A novel metal-dye detection system permits picomolar-range h.p.l.c. analysis of inositol polyphosphates from nonradioactively labeled cell or tissue specimens, *Biochem. J.*, 254, 585-591, 1988.

72. **Michell, R.H.,** Inositol phospholipids and cell surface receptor function, *Biochim. Biophys. Acta*, 415, 81-147, 1975.

73. **Michell, R.H., and Kirk, C.J.,** Why is phosphatidylinositol degraded in response to stimulation of certain receptors? *Trends Pharmacol. Sci.*, 2, 86-89, 1981.

74. **Michell, R.H., Kirk, C.J., Jones, L.M., Downes, C.P., and Creba, J.A.,** The stimulation of inositol lipid metabolism that accompanies calcium mobilization in stimulated cells: Defined characteristics and unanswered questions, *Philos. Trans. R. Soc. Lond. [Biol.], 296,* 123-137, 1981.

75. **Molina y Vedia, L.M., and Lapetina, E.G.,** Phorbol 12,13-debutyrate and 1-oleyl-2-acetyldiacylglycerol stimulate inositol trisphosphate dephosphorylation in human platelets, *J. Biol. Chem., 261,* 10493-10495, 1986.

76. **Morris, A.P., Gallacher, D.V., Irvine, R.F., and Petersen, O.H.,** Synergism of inositol trisphosphate and tetrakisphosphate in activating Ca^{2+}-dependent K^+ channels, *Nature, 330,* 653-655, 1987.

77. **O'Rourke, F.A., Halenda, S.P., Zavoico, G.B., and Feinstein, M.B.,** Inositol 1,4,5-trisphosphate releases Ca^{2+} from a Ca^{2+}-transporting membrane vesicle fraction derived from human platelets, *J. Biol. Chem., 260,* 956-962, 1985.

78. **O'Rourke, F.A., Zavoico, G.B., Smith, L.H., Jr., and Feinstein, M.B.,** Stimulus-response coupling in a cell-free platelet membrane system GTP-dependent release of Ca^{2+} by thrombin, and inhibition by pertussis toxin and a monoclonal antibody that blocks calcium release by IP_3, *FEBS Lett., 214,* 176-180, 1987.

79. **Poll, C., Kyrle, P., and Westwick, J.,** Activation of protein kinase C inhibits sodium fluoride-induced elevation of human platelet cytosolic free calcium and thromboxane B2 generation, *Biochem. Biophys. Res. Commun., 136,* 381-389, 1986.

80. **Putney, J.W., Jr.,** A model for receptor-regulated calcium entry, *Cell Calcium, 7,* 1-12, 1986.

81. **Rengasamy, A., and Feinberg, H.,** Inositol 1,4,5-trisphosphate-induced calcium release from platelet plasma membrane vesicles, *Biochem. Biophys. Res. Commun., 150,* 1021-1026, 1988.

82. **Rittenhouse, S.E., and Horne, W.C.,** Ionomycin can elevate intraplatelet Ca^{2+} and activate phospholipase A without activating phospholipase C, *Biochem. Biophys. Res. Commun., 123,* 393-397, 1984.

83. **Rittenhouse, S.E., and Sasson, J.P.,** Mass changes in myoinositol trisphosphate in human platelets stimulated by thrombin. Inhibitory effects of phorbol ester, *J. Biol. Chem., 260,* 8657-8660, 1985.

84. **Rossier, M.F., Capponi, A.M., and Vallotton, M.B.,** Metabolism of inositol 1,4,5-trisphosphate in permeabilized rat aortic smooth-muscle cells. Dependence on calcium concentration, *Biochem. J., 245,* 305-307, 1987.

85. **Rossier, M.F., Dentand, I.A., Lew, P.D., Capponi, A.M., and, Vallotton, M.B.** Interconversion of inositol (1,4,5)-trisphosphate to inositol (1,3,4,5)-tetrakisphosphate and (1,3,4)-trisphosphate in permeabilized adrenal glomerulosa cells is calcium-sensitive and ATP-dependent, *Biochem. Biophys. Res. Commun., 139,* 259-265, 1986.

86. **Sage, S.O., and Rink, T.J.,** The kinetics of changes in intracellular calcium concentration in fura-2-loaded human platelets, *J. Biol. Chem., 262,* 16364-16369, 1987.

87. **Shears, S.B., Parry, J.B., Tang, E.K., Irvine, R.F., Michell, R.H., and Kirk, C.J.,** Metabolism of D-myo-inositol 1,3,4,5-tetrakisphosphate by rat liver, including the synthesis of a novel isomer of myo-inositol tetrakisphosphate, *Biochem. J., 246,* 139-147, 1987.

88. **Shukla, S.D.,** Platelet activating factor-stimulated formation of inositol triphosphate in platelets and its regulation by various agents including Ca^{2+}, indomethacin, CV-3988, and forskolin, *Arch. Biochem. Biophys., 240,* 674-681, 1985.

89. **Siess, W., and Binder, H.,** Thrombin induces the rapid formation of inositol bisphosphate and inositol trisphosphate in human platelets, *FEBS Lett., 180,* 107-112, 1985.

90. **Siess, W., Stifel, M., Binder, H., and Weber, P.C.,** Activation of V1-receptors by vasopressin stimulates inositol phospholipid hydrolysis and arachidonate metabolism in human platelets, *Biochem. J., 233,* 83-91, 1986.

91. **Stephens, L.R., Hawkins, P.T., Barker, C.J., and Downes, P.C.,** Synthesis of myo-inositol 1,3,4,5,6-pentakisphosphate from inositol phosphates generated by receptor activation, *Biochem. J., 253,* 721-733, 1988.

92. **Stephens, L., Hawkins, P.T., Carter, N., Chahwala, S.B., Morris, A.J., Whetton, A.D., and Downes, P.C.,** L-myo-inositol 1,4,5,6-tetrakisphosphate is present in both mammalian and avian cells, *Biochem. J.,* 249, 271-282, 1988.

93. **Streb, H., Irvine, R.F., Berridge, M.J., and Schulz, I.,** Release of Ca^{2+} from a nonmitochondrial intracellular store in pancreatic acinar cells by inositol 1,4,5-trisphosphate, *Nature,* 306, 67-69, 1983.

94. **Strupish, J., Cooke, A.M., Potter, B.V.L., Gigg, R., and Nahorski, S.R.,** Stereospecific mobilization of intracellular Ca^{2+} by inositol 1,4,5-trisphosphate. Comparison with inositol 1,4,5-trisphosphorothioate and inositol 1,3,4-trisphosphate, *Biochem. J.,* 253, 901-905, 1988.

95. **Supattapone, S., Worley, P.F., Baraban, J.M., and Snyder, S.H.,** Solubilization, purification, and characterization of an inositol trisphosphate receptor, *J. Biol. Chem.,* 263, 1530-1534, 1988.

96. **Tarver, A.P., King, W.G., and Rittenhouse, S.E.,** Inositol 1,4,5-trisphosphate and inositol 1,2-cyclic 4,5-trisphosphate are minor components of total mass of inositol trisphosphate in thrombin-stimulated platelets. Rapid formation of inositol 1,3,4-trisphosphate, *J. Biol. Chem.,* 262, 17268-17271, 1987.

97. **Theibert, A.B., Supattapone, S., Worley, P.F., Baraban, J.M., Meek, J.L., and Snyder, S.H.,** Demonstration of inositol 1,3,4,5-tetrakisphosphate receptor binding, *Biochem. Biophys. Res. Commun.,* 148, 1283-1289, 1987.

98. **Tohmatsu, T., Hattori, H., Nagao, S., Ohki, K., and Nozawa, Y.,** Reversal by protein kinase C inhibitor of suppressive actions of phorbol-12-myristate-13-acetate on polyphosphoinositide metabolism and cytosolic Ca^{2+} mobilization in thrombin-stimulated human platelets, *Biochem. Biophys. Res. Commun.,* 134, 868-875, 1986.

99. **Tysnes, O-B., Verhoeven, A.J.M., and Holmsen, H.,** Phosphoinositide metabolism in resting and thrombin-stimulated human platelets, *FEBS Lett.,* 218, 68-72, 1987.

100. **Verhoeven, A.J.M., Tysnes, O.B., Aarbakke, G.M., Cook, C.A., and Holmsen, H.,.** Turnover of the phosphomonoester groups of polyphosphoinositol lipids in unstimulated human platelets, *Eur. J. Biochem.,* 166, 3-9, 1987

101. **Vickers, J.D., Kinlough-Rathbone, R.L., and Mustard, J.F.,** Accumulation of the inositol phosphates in thrombin-stimulated, washed rabbit platelets in the presence of lithium, *Biochem. J.,* 224, 399-405, 1984.

102. **Vickers, J.D., Kinlough-Rathbone, R.L., and Mustard, J.F.,** The decrease in phosphatidylinositol 4,5-bisphosphate in ADP-stimulated washed rabbit platelets is not primarily due to phospholipase C activation, *Biochem. J.,* 237, 327-332, 1986.

103. **Watson, S.P., and Lapetina, E.G.,** 1,2-Diacylglycerol and phorbol ester inhibit agonist-induced formation of inositol phosphates in human platelets: Possible implications for negative feedback regulation of inositol phospholipid hydrolysis, *Proc. Natl. Acad. Sci. USA,* 82, 2623-2626, 1985.

104. **Watson, S.P., McConnell, R.T., and Lapetina, E.G.,** The rapid formation of inositol phosphates in human platelets by thrombin is inhibited by prostacyclin, *J. Biol. Chem.,* 259, 13199-13203, 1984.

105. **Watson, S.P., McNally, J., Shipman, L.J., and Godfrey, P.P.,** The action of the protein kinase C inhibitor, staurosporine, on human platelets. Evidence against a regulatory role for protein kinase C in the formation of inositol trisphosphate by thrombin, *Biochem. J.,* 249, 345-350, 1988.

106. **Watson, S.P., Reep, B., McConnell, R.T., and Lapetina, E.G.,** Collagen stimulates [^{3}H]inositol trisphosphate formation in indomethacin-treated human platelets, *Biochem. J.,* 226, 831-837, 1985.

107. **Watson, S.P., Ruggiero, M., Abrahams, S.L., and Lapetina, E.G.,** Inositol 1,4,5-trisphosphate induces aggregation and release of 5-hydroxytryptamine from saponin-permeabilized human platelets, *J. Biol. Chem.,* 261, 5368-5372, 1986.

108. **Wiley, J.S., Quinn, M.A., and Connellan, J.M.,** Estimation of platelet size by measurement of intracellular water space using an oil technique, *Thromb. Res.,* 31, 261-268, 1983.

109. **Wilson, D.B., Bross, T.E., Sherman, W.R., Berger, R.A., and Majerus, P.W.**, Inositol cyclic phosphates are produced by cleavage of phosphatidylphosphoinositols (poly-phosphoinositides) with purified sheep seminal vesicle phospholipase C enzymes, *Proc. Natl. Acad. Sci. USA*, 82, 4013-4017, 1985.

110. **Wilson, D.B., Connolly, T.M., Bross, T.E., Majerus, P.W., Sherman, W.R., Tyler, A.N., Rubin, L.J., and Brown, J.E.**, Isolation and characterization of the inositol cyclic phosphate products of polyphosphoinositide cleavage by phospholipase C. Physiological effects in permeabilized platelets and Limulus photoreceptor cells, *J. Biol. Chem.*, 260, 13496-13501, 1985.

111. **Wilson, D.B., Connolly, T.M., Ross, T.S., Ishii, H., Bross, T.E., Deckmyn, H., Brass, L.F., and Majerus, P.W.**, Phosphoinositide metabolism in human platelets, *Adv. Prostaglandin Thromboxane Leukotriene Res.*, 17, 558-562, 1987.

112. **Wong, N.S., Barker, C.J., Shears, S.B., Kirk, C.J., and Michell, R.H.**, Inositol 1:2(cyclic), 4,5-trisphosphate is not a major product of the inositol phospholipid metabolism in vasopressin-stimulated WRK1 cells, *Biochem. J.*, 252, 1-5, 1988.

113. **Yoshida, K., Dubyak, G., and Nachmias, V.T.**, Rapid effects of phorbol ester on platelet shape change, cytoskeleton and calcium transient, *FEBS Lett.*, 206, 273-278, 1986.

Chapter 5

POLYPHOSPHOINOSITIDE METABOLISM IN RESTING AND STIMULATED PLATELETS

Holm Holmsen, Ole-Bjørn Tysnes, Adrie J.M. Verhoeven

PLATELET ACTIVATION

The individual steps in this hemostatic process have been thoroughly studied *in vitro*, and these studies have defined the platelet responses and their interactions (Figure 1). A primary agonist (e.g., collagen, ADP, thrombin) interacts with its receptor on the platelet surface, which activates polyphosphoinositide-specific phospholipase C (PPI-PLC) by mechanisms involving guanine nucleotide-binding proteins.[3] The activated PPI-PLC hydrolyzes phosphatidylinositol 4,5-bisphosphate (PIP_2) diesteratically to form inositol 1,4,5-trisphosphate (1,4,5-IP_3) and 1,2-sn-diacylglycerol (DAG). The 1,4,5-IP_3 releases Ca^{2+} from intracellular stores and this "cytoplasmic" Ca^{2+} will now promote, probably via protein phos–phorylations, the individual platelet responses. Some of the responses indicated in Figure 1 are triggered by increasing levels of cytoplasmic Ca^{2+} [9,24] in the following order: shape change, aggregation and dense granule secretion (i.e., secretion of serotonin and ATP). The DAG formed in the initial PPI-PLC reaction activates protein kinase C (PKC) which enhances these Ca^{2+}-induced platelet responses,[22,23] possibly by cytoplasmic alkalinization through a Na+/H+ exchange reaction.[26] The other responses shown in Figure 1, α-granule secretion, arachidonate liberation and acid hydrolase secretion, are also triggered by cytoplasmic Ca^{2+},[9] but it is not known whether they have threshold Ca^{2+} concentrations and are enhanced by PKC activation.

An important feature in platelet activation is positive feedback, i.e., enhancement of the activation by platelet-produced agonists, such as secreted ADP and serotonin, and prostaglandins/thromboxane A_2 derived from the liberated arachidonate (Figure 1). The platelet-produced agonists interact with each other and the primary agonist in a synergistic manner, thus making the resulting degree of response greater than the sum of the responses caused by the individual agonists.[27] One assumes that platelet activation *in vivo* is a result of simultaneous stimulation by many agonists derived from the subendothelium (collagen), platelets (ADP, serotonin, prostaglandins, thromboxane A_2) and plasma (thrombin, adrenaline, vasopressin) and working in synergy. The mechanism(s) for synergistic activation are not clear but may be located in the signal transduction cascade at the level of PPI-PLC activation, at least for thrombin and adrenaline.[4,28]

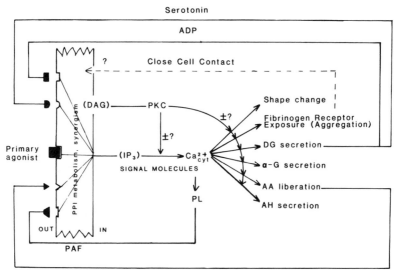

FIGURE 1. Schematic overview of platelet activation, responses and positive feedback. See text for explanation.

POLYPHOSPHOINOSITIDE (PPI) CYCLE AND RADIOISOTOPIC LABELING

This chapter describes the metabolism of the PPIs in both resting and agonist-activated platelets. The further metabolism of the inositol phosphate produced from the PPI and their role in Ca^{2+} mobilization were previously discussed in this book (see chapter 4 by James L. Daniel).

Figure 2 demonstrates the cyclic metabolism that the PPIs undergo in stimulated platelets (the "PPI cycle") and provides the formulas for the metabolites involved. Receptor occupancy leads to PPI-PLC activation and diesteratic cleavage of PIP_2 (reaction 3). While the 1,4,5-IP_3 formed is metabolized along several routes to free inositol, most of the DAG formed is phosphorylated by ATP to phosphatidic acid (PA)[20] (reaction 6). PA, which is the only metabolite to accumulate (see below), is converted to cytidine-5'-diphosphatediacylglycerol (CDPDAG) (reaction 8) and this activated form of PA will be esterified with the 1-hydroxyl by free inositol, thus forming phosphatidylinositol (PI) (reaction 9). PI is phosphorylated by ATP at the 4-hydroxyl in the inositol ring to phosphatidylinositol 4-phosphate (PIP) (reaction 1), which is in turn phosphorylated by ATP at the 5-hydroxyl in the inositol ring to PIP_2 (reaction 2) by appropriate kinases. In the resting and activated platelets, the 4- and 5-phosphoryl groups are also rapidly removed by appropriate phosphohydrolases

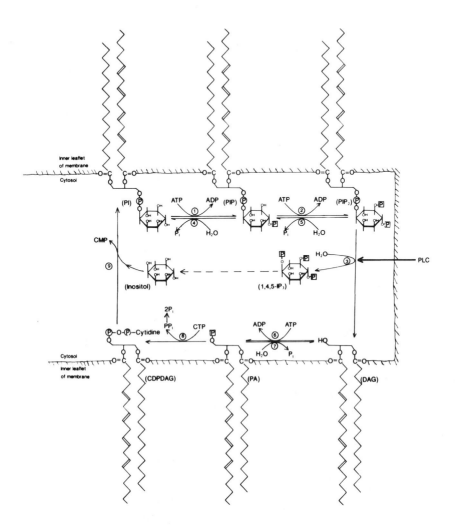

FIGURE 2. The polyphosphoinositide (PPI) cycle in platelets. The glycerol backbone of PPIs and their metabolites contain stearoyl and arachidonyl in the 1- and 2-position, respectively. These acyl moieties span the inner leaflet of the plasma membrane by which the individual metabolites are anchored to the cytoplasmic side of this membrane. The hydrophilic part of the metabolites probably reach out in the cytoplasm where their inositol and glycerol (3-) hydroxyls are undergoing the phosphorylations and dephosphorylations shown. Phosphomono- and diester groups are shown as P within squares and circles, respectively. Note that the phosphomonoester in PA becomes the phosphodiester in PI, PIP and PIP_2, and that the conversion of 1,4,5-IP_3 to free inositol (broken line) are given without details. The scheme is based on literature reviewed by Holmsen[10] and Vickers and Mustard.[38] The individual reactions are catalyzed by the following enzymes: 1 = PI kinase (EC 2.7.1.67); 2 = PIP kinase (EC 2.7.1.68); 3 = PIP_2 phosphodiesterase (EC 3.1.4.11); 4 = PIP phosphomonoesterase (EC not recorded); 5 = PIP_2 phosphomonoesterase (EC 3.1.3.36); 6 = DAG kinase (EC 2.7.1.107); 7 = PA phosphatase (EC 3.1.3.4); 8 = PA cytidyltransferase (EC 2.7.7.41); 9 = CDPDAG-inositol 3-phosphatidyltransferase (EC 2.7.8.11).

(reactions 4 and 5). Because these kinases and phosphohydrolases are simultaneously active, the phosphoryl groups go on and off the 4- and 5-hydroxyls groups. This results in a net consumption of ATP in apparently futile reaction cycles. The levels of PIP and PIP_2 are, therefore, typical steady-state levels (see below).

Human platelets contain about 1400, 170 and 130 nmoles/10 platelets of PI, PIP and PIP_2, respectively.[30,35] These amounts of PIP and PIP_2 are so small that their measurements by phosphorus determination are very time- and platelet-consuming. In most experiments on the metabolism of the PPIs and their role in signal transduction, the PPIs have, therefore, been prelabeled by incubation of the platelets with radioactive orthophosphate,[1,12,35] arachidonate,[11] glycerol[30] or inositol.[15]

Orthophosphate is incorporated as the γ-phosphoryl group of ATP into two different phosphate bonds, i.e., as monoesters in the 4- and 5-positions in PIP and PIP_2 and as a diester between C_3 in glycerol and C_1 in inositol in PI, PIP and PIP_2. The degree of radiolabeling is very different as the monoester phosphates are labeled at the same specific radioactivity as the γ-phosphoryl of ATP, while the diester has only 1/30 - 1/20 of the ATP radioactivity (see below). This means that ^{32}P-labeling results in stronger labeling of PIP and PIP_2 than of any other platelet phospholipid (Figure 3) and represents the most sensitive method for monitoring the steady-state levels of these PPIs (see below). The PPI cycle consumes at least three ATP molecules directly (reactions 1, 2 and 6 in Figure 2), and two molecules of ATP indirectly in reaction 8 to regenerate cytidine-5'-triphosphate (CTP) from the cylidine-5'-monophosphate produced in reaction 9, suggesting a total expenditure of 5 ATP molecules per cycle. The ATP expenditure is, however, far greater because of the (unknown) ATP usage in the two futile reaction cycles discussed above.

Glycerol is incorporated into the PPI cycle as PA after it has been phosphorylated (by ATP) in the 3-position and acylated in the 1- and 2-positions.[32] However, the fatty acids in PI, PIP and PIP_2 are very conserved as the 1-stearoyl, 2-arachidonyl species constitutes more than 90% of these inositol lipids, whereas the PA in resting platelets only has 50% of this species.[20] It is therefore not clear whether the PPIs newly formed from PA have the stearoyl/arachidonyl composition, and thus whether glycerol initially labels representative inositol lipids. However, we have shown that radioactive glycerol eventually becomes distributed among PI, PIP and PIP_2 according to their masses.[33]

The mechanisms for incorporation of radioactive arachidonate are not entirely clear. Some arachidonate, but not all (see above), may be incorporated into the DAG moiety of PI, PIP and PIP_2 during *de novo* PA formation. Possibly, a substantial proportion of arachidonate is incorporated into these lipids by remodeling reactions.[21] It is not known whether radioactive arachidonate distributes among the inositol lipids according to their masses.

Radioactive inositol is presumably entering the inositol lipids through reaction 9 in Figure 2, a reaction which is promoted by Mn^{2+} ions.[5,19] However,

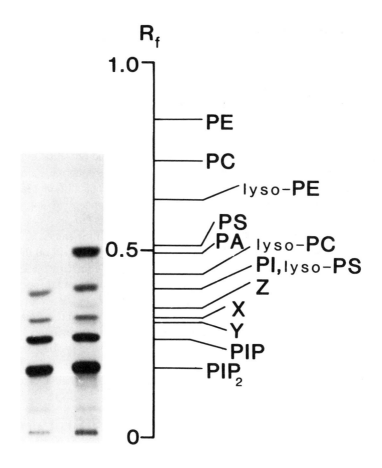

FIGURE 3. ^{32}P-labeled phospholipids from resting and thrombin-treated platelets. Human platelet-rich plasma was incubated with ^{32}P-P$_i$ for 60 min and plasma with remaining, unincorporated ^{32}P-P$_i$ removed by gel filtration. Portions of the platelet suspension were incubated for 60 sec at 37°C with 0.5 U/ml of thrombin or with 0.15 M NaCl. Chloroform-methanol extracts were prepared and separated on oxalated silica gel plates with chloroform/methanol/20% methylamine in water (60:35:10). The radioactive fractions were visualized by autoradiography (left lane = control platelets; right lane = thrombin-treated platelets). Further details are described elsewhere.[29] (Reprinted with permission from *Thromb. Res.* 40, copyright 1985, Pergamon Press PLC.)

inositol is a very impermeable molecule and its entry into platelets would therefore depend greatly on whether the platelet membrane of individual cells in a given platelet suspension remains intact. The impermeability of the platelet and the lack of data regarding possible effects of Mn^{2+} on platelet responsivity make the use of platelet labeling with radioactive inositol a questionable procedure for studying the involvement of inositol lipids and phosphates in platelet signal transduction.

TIME COURSES OF CHANGES IN ^{32}P-PPIs DURING PLATELET STIMULATION

It follows from the above that incubation of platelets with ^{32}P-orthophosphate gives strong labeling in the PPIs. Figure 3 shows an autoradiogram of acid lipid extracts of resting and stimulated ^{32}P-labeled platelets separated by thin layer chromatography (TLC). PI, the PPIs and an unknown substance (X) are strongly labeled, whereas there is hardly any labeling of the major phospholipids phosphatidylcholine, phosphatidylethanolamine and phosphatidylserine in the resting platelets. On the other hand, these major phospholipids are strongly labeled when platelets are incubated with ^{3}H-glycerol. This different labeling of these phospholipids with ^{3}H-glycerol and ^{32}P-orthophosphate suggests that glycerol is not incorporated by mass *de novo* synthesis.[32]

Stimulation of the platelets with 0.5 U/ml thrombin causes a dramatic increase in the ^{32}P-content in PA and a slight increase in ^{32}P-PI on the autoradiogram (Figure 3). Quantitative analyses of the time course of these changes show that thrombin causes an increase in the radioactivity of all four metabolites in platelets shortly after the end of (pre)incubation with ^{32}P-orthophosphate (Figure 4A), while it causes a decrease in ^{32}P-PI and the same increase in the other three metabolites when added 90 min after the end of labeling (Figure 4B). This difference in the behavior of ^{32}P-PI and the increase in radioactive PIP and PIP$_2$, which actually are consumed in the process, represent basic problems for interpretation of results obtained with ^{32}P-labeling, which will be discussed in detail.

TURNOVER OF PI IN RESTING AND STIMULATED PLATELETS

The mass of PI decreases during the thrombin-platelet interaction (Table 1) with a time course independent of the time after end of prelabeling,[31] and the time course for PI mass is superimposable on the changes in ^{32}P-radioactivity 90 min after the end of prelabeling (Figure 4B). However, while the specific radioactivity in PIP and PIP$_2$ decreased slightly with time after the end of incubation, that in PI increased 3-4 times during the first 90 min after the end of labeling with ^{32}P-orthophosphate (Table 2). Even after this increase, the specific radioactivity in the diester phosphate of PI is only 17% of that in the γ-phosphate in ATP, while the monoesters in PIP and PIP$_2$ have the same specific radioactivity as the γ-phosphate at all times after the end of prelabeling (Table 2).

From the data in Table 2 it follows that the PI diester phosphate is not in equilibrium with ATP and has a turnover of 2.4 nmol/min/10^{11} cells in resting platelets (see illustration page 112). The marked increase in the PI-radioactivity in thrombin-stimulated platelets right at the end of labeling (Figure 4A), while PI mass is decreasing and the specific radioactivity is only 4% of that in ATP

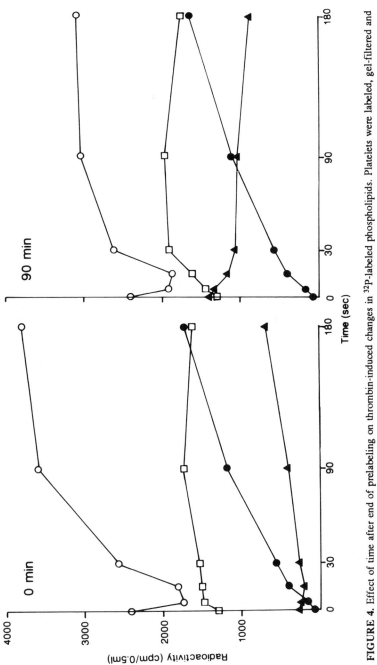

FIGURE 4. Effect of time after end of prelabeling on thrombin-induced changes in ^{32}P-labeled phospholipids. Platelets were labeled, gel-filtered and incubated as outlined in Figure 3, except that lipid extracts were made from aliquots of the incubation mixture at the times shown, and the incubations were done immediately (A) and 90 min after (B) the platelets had been gel-filtered. The data are reconstructed from Tysnes *et al.*[31] ○ = PIP$_2$; □ = PIP; ● = PA; ▲ = PI.

TABLE 1
Specific Radioactivity of the Mono- and Diester Phosphates in Phosphoinositides of Stimulated Platelets.

Experiments were carried out as in Figure 4A and analyzed for mass (nmoles/10^{11} platelets) and specific radioactivity (cpm/nmole) of the mono- and diester phosphates of PI, PIP and PIP$_2$.[30]

		\multicolumn{4}{c}{Incubation time with thrombin (sec)}			
		0	15	60	120
PI	Mass	1408	1238	1140	921
	Sp. Ra.	570	710	1120	2480
PIP	Mass	168	167	219	205
	Sp. ra., monoest.	14000	12900	12300	11500
	Sp. ra., diest.	550	670	1070	2470
PIP$_2$	Mass	132	148	172	173
	Sp. ra., monoest.	14100	14350	13600	12500
	Sp. ra., diest.	560	780	1110	2840

TABLE 2
Effect of Time After End of Prelabeling with ^{32}P-P$_i$ on the Relative Specific Radioactivity of Phosphates in ATP, ADP and Phosphoinositides in Resting Platelets.

The experiment is similar to that described in Figure 3, and the numbers shown (specific ^{32}P-radioactivity relative to that of the ATP γ-phosphate just after end of labeling) are composed from results from Tysnes *et al.*,[30,31] Verhoeven *et al.*[35]

Substance	Type of phosphate bond	\multicolumn{2}{c}{Time after end of labeling}	
		0	90'
ATP	Anhydride (β and γ)	100.0	78.3
ADP	Anhydride (β)	110.0	—
PI	Diester	3.2	16.0
PIP	Monoester	90.0	84.5
	Diester	3.1	18.5
PIP$_2$	Monoester	90.0	86.4
	Diester	3.6	18.5

(Table 2), means that PI of much higher specific radioactivity is being made because of increased turnover during the thrombin-platelet interaction. With 0.5 U/ml of thrombin, this increased turnover has been calculated to be 20 nmol/min/10^{11} platelets (see illustration page 112), so that stimulation with thrombin causes an 8- to 9-fold stimulation of PI turnover in platelets. When the ^{32}P-labeled platelets are stimulated 90 min after the end of prelabeling, i.e., when the specific PI radioactivity is about 20% of that in ATP, the total PI radioactivity decreases (Figure 4B) almost in parallel with the mass, because the change in specific radioactivity now is very small.

TURNOVER OF PIP AND PIP$_2$ IN RESTING AND STIMULATED PLATELETS

Since the monoester phosphates of the PPIs have the same specific radioactivity as the γ-phosphate in ATP (Table 2), the turnover of these monoester phosphates is equal to or greater than that of ATP. By determining the distribution of ^{32}P-radioactivity during uptake of ^{32}P-orthophosphate in platelets, so-called "non-equilibrium" labeling, 7% of the ^{32}P-radioactivity was confined to the PPIs at any stage.[35] Since the turnover of the monoester phosphates in the PPIs and the β- and γ-phosphates in ATP is the same, the ^{32}P should distribute among these phosphates according to the pool sizes. These are 424 and 5900 nmoles/10^{11} for the PPIs (160 + 2 × 132, see above) and metabolic ATP+ ADP,[14] respectively, which gives a 7.2:100 ratio, and is, thus, very close to the value obtained from the uptake studies. These observations strongly suggest that as much as 7% of the ATP produced in resting platelets is consumed to maintain the PPIs in the phosphorylated state. The production of ATP in 10^{11} resting platelets is 3000 nmoles/min; hence, the fraction of energy consumed in phosphorylation/dephosphorylation of the PPI monoesters is at least 210 nmoles/ min/10^{11} platelets (see page 112). With 424 nmoles PPI monoester/10^{11} platelets, the turnover rate of the monoesters is at least 0.49 min^{-1} or 0.008 s^{-1}.

These data demonstrate that the monoester phosphates of the PPIs are in metabolic equilibrium with the high-energy phosphates in ATP and ADP. The specific radioactivity of the diester phosphate in the PPIs was found to be practically identical to that of PI at any stage during both storage (Table 2) and treatment with thrombin (Table 1) of ^{32}P-prelabeled platelets. This strongly suggests that PI is in metabolic equilibrium with both PIP and PIP$_2$ in our studies, i.e., that there is no compartmentalization of the phosphoinositides. Vickers and Mustard[37] reported compartmentalization of platelet PPIs based on changes in ^{32}P-labeling by storage; however, they did not distinguish between mono- and diester phosphates, which clearly is essential.

The specific ^{32}P-radioactivity of the monoester phosphates in PIP and PIP$_2$ is unchanged during both storage (Table 2) and treatment with thrombin (Table 1) of ^{32}P-prelabeled platelets. Since the specific radioactivities of the monoester phosphates are 20-30 times that of the diester phosphate (Tables 1 and 2), the total radioactivities of PIP and PIP$_2$ is a sensitive and fair (relative) measure of their masses. It follows, then, from Figures 4A and 4B that, except for a transient decrease in the PIP$_2$-radioactivity, the general trend is that the masses of PIP and PIP$_2$ increase during the thrombin-platelet interaction. Such an increase seems to contradict our claim above and illustrated in Figure 2 that PIP$_2$ is consumed in the PPI-PLC reaction during platelet activation. This apparent contradiction is easily explained when the rates of essential steps in the PPI cycle are known; this will be discussed below.

RAPID INFLUX AND EFFLUX OF INORGANIC PHOSPHATE DURING PLATELET STIMULATION

In all experiments discussed above (e.g., Figures 3 and 4, Tables 1 and 2) the platelets had been incubated with ^{32}P-inorganic orthophosphate (P_i) in platelet-rich plasma for 1 hr at 37°C, followed by removal of excess (non-incorporated) ^{32}P-P_i by transferring the platelets by gel filtration into a P_i-free Tyrode's solution. The rationale behind this procedure is to label the metabolic pool of ATP and those metabolites that turn over rapidly, and then ascertain that no further exchange with P_i takes place in these molecules during subsequent treatment of the platelets (so-called "equilibrium labeling"). As discussed above, this approach makes it possible to monitor changes in the PPI masses as changes in their total radioactivity.

The experiments do not, however, tell us anything about the turnover of the PPIs in thrombin-stimulated platelets. In an attempt to determine this turnover, Wilson *et al.*[40] stimulated platelets with thrombin during uptake of ^{32}P-P_i. They demonstrated a massive increase in the total radioactivity of PIP and PIP_2 which was interpreted to mean an increase in turnover. However, stimulation of platelets with thrombin under these "non-equilibrium" conditions causes a rapid increase in the uptake of ^{32}P-P_i by the platelets (Figure 5). This burst in ^{32}P-P_i uptake caused an immediate and identical increase in the specific radioactivity of ATP, PIP and PIP_2.[36] This identical increase occurs because the turnover of the terminal phosphate in ATP, and hence of the phosphomonoesters in PIP and PIP_2 (above), is more than 200 times greater than the rate of P_i uptake in resting platelets.[36] Thus, when the rate of uptake increases 5-10 times by thrombin stimulation (Figure 5), this rate is still 20-40 times smaller than the turnover rate of the rapidly exchanging phosphate moieties in ATP, PIP and PIP_2 in resting platelets. The turnover of ATP increases 2-3 times during platelet stimulation,[34] but the approach with "non-equilibrium" labeling does not give any information about whether the turnover of the phosphomonoesters in PIP and PIP_2 increases in the same or greater manner.

Our inability to determine the turnover of PIP and PIP_2 by the (simultaneous) ^{32}P labeling technique is due to the tight metabolic equilibrium between the γ-phosphoryl group of ATP and the monoester phosphates of PIP and PIP_2 in stimulated (Table 1) and resting (Table 2) platelets. Also, the metabolic homogeneity among PI, PIP and PIP_2, conferred by the specific radioactivity of the diester phosphate (Tables 1 and 2) makes it impossible to use this parameter for their turnover determination. Only the possible existence of a metabolic non-equilibrium condition would enable us to determine whether stimulation of platelets leads to changes in the turnover of the PPIs; this condition is yet to be found.

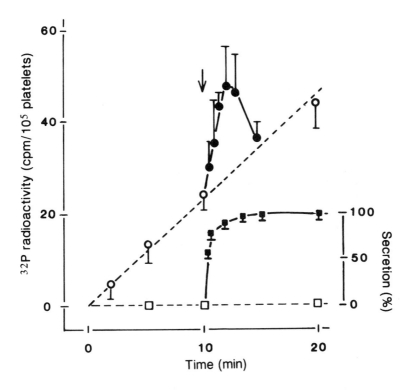

FIGURE 5. Effect of thrombin stimulation on phosphate uptake by platelets. Gel-filtered (non-labeled) platelets were incubated with 420 μM ^{32}P-P$_i$ at 37°C. Thrombin (f.c. 0.5 U/ml) was added to an aliquot of the incubation mixture 10 min after start of incubation. Samples were removed from both incubation mixtures at the times shown, and the amount of platelet-bound ^{32}P (circles) was determined after the cells were isolated by centrifugation through an oil layer (left abscissa). The supernatant above the oil layer was analyzed for amounts of ATP+ADP (squares) in order to determine the degree of dense granule secretion (right abscissa). The filled and open symbols refer to presence and absence of thrombin, respectively. For further details, see Verhoeven *et al.*[36] (Reprinted with permission from *J. Biol. Chem.* **262**, copyright 1987, American Society for Biochemistry and Molecular Biology.)

RATES OF PRODUCTION AND CONSUMPTION OF PA IN STIMULATED PLATELETS—THE EXISTENCE OF A NON-EQUILIBRIUM CONDITION

In contrast to the metabolic equilibrium situation discussed above for the PPIs, resting platelets contain PA which has a much lower specific radioactivity than the γ-phosphoryl group of ATP when the platelets have been prelabeled with ^{32}P-P$_i$ (Table 3). Thus, just as for PI above, we also have a non-equilibrium condition for PA in resting platelets. Stimulation of such platelets leads to synthesis of considerable amounts of PA with the same specific radioactivity as ATP. In this way the specific radioactivity of bulk PA (the PA extracted directly from the platelets) increases upon stimulation of the platelets with thrombin, and it approaches that of ATP (Table 3).

Stimulation of ^{3}H-glycerol-prelabeled platelets with thrombin caused ^{3}H-PA to accumulate at a constant rate, suggesting a linear synthesis and removal of PA in (maximally) stimulated platelets.[33] This linearity enabled us to develop a mathematical model based on conservation of mass and ^{32}P-radioactivity and the assumption (above) that the phosphate in newly synthesized PA molecules has the same specific radioactivity as the γ-phosphoryl group in ATP.[33] With this model and the values shown in Table 3, it can be calculated that the rate of the thrombin-induced synthesis and removal of PA is 107 and 52 nmol/min/ 10^{11}, respectively. We believe that in stimulated platelets, the major processes for PA synthesis and removal are the DAG kinase and the PA:CTP cytidyl transferase reactions, respectively (see illustration page 112).

METABOLIC HETEROGENEITY OF PA IN RESTING AND STIMULATED PLATELETS

Two-dimensional TLC of lipid extracts from unstimulated, ^{32}P-prelabeled platelets always produced a tadpole-shaped spot of PA in the second direction (Figure 6, left). Since this direction separates primarily according to mass, it is reasonable to assume that the "tail" component represents PA species with lower MW fatty acids than those of the "head" component. We have seen above (Figures 3 and 4) that stimulation of platelets leads to an enormous accumulation of ^{32}P-labeled PA separated by one-dimensional TLC. With the two-dimensional procedure, it is apparent that this increase is mainly confined to the head fraction of PA (Figure 6, right). This is also shown quantitatively in Figure 7 where the head and tail fractions were scraped off the chromatograms and counted.

Preliminary experiments suggest that the two PA fractions have different metabolic roles. When platelets were prelabeled with both ^{32}P-P$_i$ and ^{3}H-glycerol, the content of ^{32}P and ^{3}H in the head fraction increased by 30 and 1-2 times, respectively, by thrombin treatment, while the increases of ^{32}P and ^{3}H were similar in the tail fraction (Figure 7). Thus, the major part of PA synthesized during platelet stimulation (head fraction) is apparently much poorer in the ^{3}H-glycerol label than that of the minor tail fraction. One may speculate that the PA head fraction is derived exclusively from the PPI cycle (Figure 2; Figure 8 below), while the PA tail fraction is representative of (increased) *de novo* phospholipid synthesis which produces PA with fatty acids added at random. Since the DAG backbone of the PPIs conserves stearoyl and arachidonyl residues in the 1- and 2-*sn* position, respectively,[20] fatty acid analysis of the PA head and tail fractions is necessary in order to further elucidate this problem.

We discussed above that the much lower specific radioactivity of PA than of ATP in ^{32}P-labeled, resting platelets (Table 3) could mean that their (bulk) PA is turning over very slowly. However, the heterogeneity of PA shown in Figure

TABLE 3.
Specific Radioactivity of PA in Resting and Stimulated ^{32}P-Labeled Platelets.

[32]P-labeled, gel-filtered platelets were incubated at 37°C for 180 sec with 0.5 U/ml of thrombin (stim) or 0.15 M NaCl (contr) and the amount of ^{32}P and mass of PA and ATP were determined. The numbers represent six individual experiments.[33]

Compound	nmol/10[11] cells	cpm/nmole
PA (contr)	171 ± 25	442 ± 42
PA (stim)	336 ± 36	7044 ± 395
ATP		9502 ± 1037

6 offers the possibility that one fraction (head) has a fast turnover (its phosphate could even be in metabolic equilibrium with ATP), while the other fraction (tail) turns over slowly, if at all. If so, the head fraction would represent PA formed in the PPI cycle in resting platelets, while the tail fraction would be PA formed by other mechanisms (phospholipase D; *de novo* synthesis).

CONVERSION RATES IN THE INDIVIDUAL STEPS IN THE PPI CYCLE OF RESTING AND STIMULATED PLATELETS—PURPOSE OF THE HIGH TURNOVER RATES OF THE PPI MONOESTERS

Figure 8 provides the turnover rates in the individual steps of the PPI cycle in resting and (thrombin) stimulated platelets. As has been pointed out, the turnover rates of the mono- and diester phosphates are very different. Since the specific radioactivity of the PPI monoester phosphates is identical to that of the γ-phosphate of ATP, these monoesters must turn over as fast or faster than the γ-phosphate of ATP in resting platelets. This turnover of ATP is about 3000 nmoles/min/10[11] platelets,[2] and the top of the cycle in Figure 8 indicates that the monoester phosphates of PIP and PIP_2 have this rapid turnover (for further discussion of this turnover, see below). Stimulation of platelets causes a rapid, twofold increase in ATP turnover,[2,34] and it is possible that the turnover of the PPI monoester phosphates increase similarly on stimulation, as indicated in Figure 8. Note that such an increase in the turnover of the monoesters does not reflect a possible increase in the net flux of inositol lipid in the PI -> PIP -> PIP_2 sequence in Figure 8 (reactions 1 and 2, Figure 3).

The PI formation rate in resting platelets and the increase in both this rate and the rates of PA formation and consumption by platelet stimulation are based on large differences in specific ^{32}P-radioactivity of the lipid-bound phosphate and the γ-phosphate in ATP. There are at least two molecular species of PA in platelets (Figure 6) that may represent PPI-related and PPI-unrelated species, and bulk PA does not change its measurable specific ^{32}P-radioactivity after the end of labeling (see zero-times of PA at 0 and 90 min, Figure 4). These facts have made it impossible to determine the turnover of PPI-related PA in resting

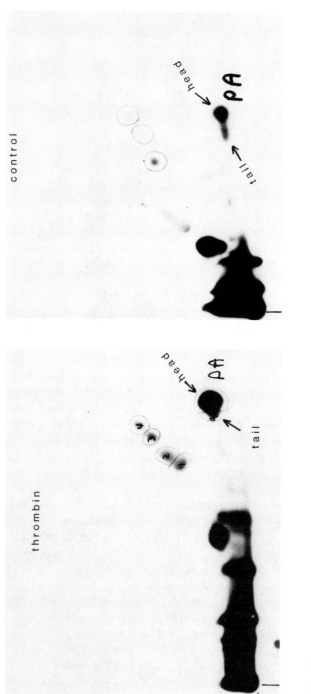

FIGURE 6. Double spot of PA. Platelets were labeled, gel-filtered, incubated with and without thrombin and extracted as outlined in Figure 3. The extracts were separated by two-dimensional chromatography on silica plates with chloroform/methanol/7 N ammonia (65:20:4) in the first direction (upward) and chloroform/acetone/methanol/acetic acid/water (50:20:10:5:5) in the second direction (from left to right). The autoradiograms shown are from separations of extracts from control (left) and thrombin-treated (right) platelets.

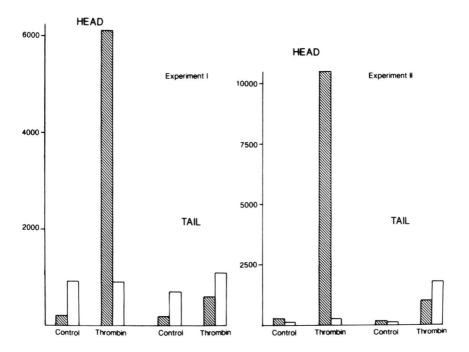

FIGURE 7. Analysis of the content of ^{3}H-glycerol and ^{32}P-P$_i$ in the head and tail fraction of the PA spot in control and thrombin-treated platelets. Platelets were labeled with ^{32}P-P$_i$ and ^{3}H-glycerol as described elsewhere,[32] stimulated with and without thrombin as described in Figure 3, and the phospholipids separated as described in Figure 6. The head and tail fractions of the PA spots were scraped off the chromatograms and their content of ^{3}H (open bars) and ^{32}P (hatched bars) determined by scintillation counting. The results are preliminary and representative of two independent experiments.

platelets by current methodologies. Since the PA level remains constant in resting platelets, we have assumed that its turnover equals that of PI, i.e., 2.4 nmoles/min/10^{11} platelets, which is used as the value for its formation and consumption in Figure 8.

It was possible to measure formation rates of PI and PA as well as PA consumption in stimulated platelets. The values are shown in Figure 8, and since we have been unable to detect any measurable increase in DAG, the rate of DAG formation (rate of the PLC reaction, number 3 in Figure 3) has been set equal to that of PA formation in Figure 8. With these assumptions, it follows that maximal stimulation of platelets leads to an almost 50-fold increase in the formation of PA, and that PA must accumulate (see Figure 3) since its rate of consumption is half that of its formation (Figure 8).

The rate of interconversion among PI, PIP and PIP$_2$ is 30-60 times that of PIP$_2$ consumption in stimulated platelets (Figure 8). The renewal rate of PIP$_2$ (and of PIP) is so much greater than that of its diesteratic hydrolysis that the diesteratic hydrolysis does not result in decrease in the PIP$_2$ level. On the

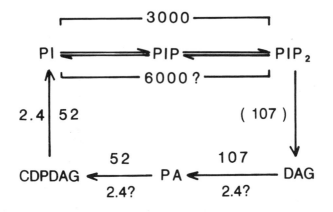

FIGURE 8. Turnover rates in individual steps of the PPI cycle in resting and stimulated platelets. The figure represents the PPI cycle shown in Figure 3 without exact chemical formulas. The numbers outside and inside the cycle represent the rates (nmoles/min/10^{11} platelets) in resting and stimulated cells, respectively. See text for further explanation.

contrary, the levels of PIP and PIP_2 actually increase (Table 1, Figure 3). However, the level of the inositol lipid pool (PI + PIP + PIP_2) decreases, with the decrease confined only to PI (Table 1). There is complete metabolic equilibrium between the diesters, the monoesters and the glycerol moieties of PI, PIP and PIP_2, and their interconversion rate is greater than the rate of draining the inositol lipid pool (see above). One would therefore expect that the mass equilibrium remained the same during stimulation, but it does not, as the (PIP + PIP_2)/PI ratio increases from 0.2 to 0.4 (Table 1). Therefore, stimulation must lead to alternation in the interconversion reactions, i.e., reactions 1,2,4 and 5 in Figure 2.

The much higher turnover of the PPIs as compared to the consumption of PIP_2 in responding platelets ensures that the cells are not depleted for this PLC substrate during stimulation. It is more difficult to understand the high turnover of the PPIs in the resting platelet and why it uses as much as 7% of the ATP produced to maintain PIP and PIP_2 in the phosphorylated state. The situation may be analogous to the treadmilling of actin in the resting platelet where F-actin is depolymerized and polymerized continuously at the expenditure of up to 40% of the ATP produced.[6] Thus, both PPI dephosphorylation/phosphorylation and actin depolymerization/polymerization are processes of high (energy) priority in the resting platelet, which might appear as futile cycles with no purpose other than producing heat.

These two "recycling" processes have in common that their forward reactions (phosphorylation, polymerization) participate actively in the signal processing machinery in stimulated platelets: PIP_2 is rapidly resynthesized from PIP and PI (Figure 2), while actin is rapidly polymerized to long fibrils beneath the plasma membrane.[7] For the stimulated platelet it is obvious that these processes must occur rapidly in order to execute signal processing as efficiently as the platelet does (shape change takes 2-3 sec, secretion starts after 5-7 sec). The possibility therefore exists that the rapid recycling of PPI and actin in the resting state is part of the cells' responsivity—processes that serve to maintain a high degree of preparedness in the cell so that a stimulus can be processed immediately. Since parts of these recycling reactions are used during signal processing, the reaction need not start from zero but is effectuated by changing the velocity of direction of an ongoing reaction. The latter is obviously more efficient than the former, and could be likened to starting movement of a car with the motor on and releasing the clutch rather than starting the movement and motor simultaneously.

For the inositol lipids in particular, presence of none, one or two monoester phosphates on the inositol ring sticking out in the cytoplasm (see Figure 2) should make dramatic changes in the charge of this part of the molecule, changes that might even lead to increased lateral movement of the PI moiety within the inner leaflet of the plasma membrane. This could increase the chemical activity of PIP_2 in the membrane and thus increase the chances for collision of PIP_2 with (activated) PLC in the membrane in a stimulated cell.

In this chapter the terms PIP and PIP_2 have been used for the PI(4)P and PI(4,5)P_2 isomers. However, the PI(3)P and PI(3,4)P_2 isomers in addition to PI(3,4,5)P_3 have recently been demonstrated in several cells, including platelets, where each 3-phosphorylated isomers constitute probably less than 1% of the amount of PI(4,5)P_2. The function of these novel polyphosphoinositides is presently unknown. A 3-kinase has been characterized indicating that the 3-phosphoinositides are produced from PI, PIP(4)P and PI(4,5)P_2.[39] Recently Cunningham and Majerus[17] claimed that PI(3)P is initially formed from PI and that this PI(3)P is consequently phosphorylated to PI(3,4)P_2 and PI(3,4,5)P_3. These 3-phosphorylated compounds are poor substrates for phospholipase C,[25] and a role in the second messenger production has therefore been difficult to envisage. Moreover, the water-soluble compounds that would result from diesteratic hydrolysis of the 3-phosphorylated phosphoinositides are all present in cells and are generally believed to be produced by phosphorylation and dephosphorylation of inositol(1,4,5)P_3 secondary to the phosphodiesteratic cleavage of the "common" PI(3,4)P_2.

Although present in small amounts in resting platelets, there is a considerable increase in PI(3,4)P_2 and PI(3,4,5)P_3 during stimulation of these cells with thrombin or U46619(1), which suggests a role in signal transduction. It has been proposed that these lipids my be involved in the actin polymerization process by causing the dissociation of the gelsolin-actin and profilin-actin complexes.[18] This has been shown in platelets for PI(4,5)P_2 at concentrations which are relevant in the context of the 3-phosphorylated compounds.[16] Moreover, actin polymerization and depolymerization in neutrophils has been shown to parallel the rise and fall in PI(3,4,5)P_3, and reversion of the agonist binding in these cells simultaneously decreases actin-polymerization and the level of PI(3,4,5)P_3.[8]

This could represent a role for this 3-phosphorylated compound, especially since PI(3,4,5)P$_3$ is the only phosphatidylinositol that is absent before stimulation and that rises transiently upon stimulation of the cells.

Certainly many aspects around the 3-phosphorylated phospho-inositides need to be clarified. Although produced upon platelet stimulation, it presently seems unlikely that they are involved in the bifurcating signal-transduction pathway initiated by phospholipase C.

ACKNOWLEDGMENTS

This work was supported by the Royal Norwegian Council for Scientific and Industrial Research, the Norwegian Society for Fighting Cancer, the Norwegian Council for Cardiovascular Research and the Family Blix' Fund.

List of Abbreviations
1,4,5-IP$_3$ = Inositol 1,4,5-trisphosphate
CDPDAG = Cytidine-5'-diphosphatediacylglycerol
CTP = Cytidine-5'-triphosphate
DAG = 1,2-sn-diacylglycerol
PA = Phosphatidic acid
P$_i$ = Inorganic phosphate
PI = Phosphatidylinositol
PIP = Phosphatidylinositol 4-phosphate
PIP$_2$ = Phosphatidylinositol 4,5-bisphosphate
PKC = Protein kinase C
PLC = Phospholipase C
PPI(s) = Polyphosphoinositide(s)
TLC = Thin layer chromatography

REFERENCES

1. **Agranoff, B. W., Murthy, P., and Seguin, E. B.,** Thrombin-induced phosphodiesteratic cleavage of phosphatidylinositol bisphosphate in human platelets, *J. Biol. Chem.*, 258, 2076-2078, 1983.

2. **Akkerman, J.W.N., Gorter, G., Schrama, L., and Holmsen, H.,** A novel technique for rapid determination of energy consumption in platelets. Demonstration of different energy consumption associated with three secretory responses, *Biochem. J.*, 210, 145-155, 1983.

3. **Brass, L.F., Shaller, C.C., and Belmonte, E.J.,** Inositol 1,4,5-trisphosphate-induced granule secretion in platelets. Evidence that the activation of phospholipase C mediated by platelet thromboxane receptors involves a guanine binding protein-dependent mechanism distinct from that of thrombin, *J. Clin. Invest.*, 79, 1269-1275, 1987.

4. **Crouch, M.F., and Lapetina, E.G.,** A role for Gi in the control of thrombin receptor-phospholipase C coupling in human platelets, *J. Biol. Chem.*, 263, 3363-3371, 1988.

5. **Culty, M., Davidson, M.M.L., and Haslam, R.J.,** Effects of guanosine 5'[γ-thio]triphosphate and thrombin on the phosphoinositide metabolism of electropermeabilized human platelets, *Eur. J. Biochem.*, 171, 523-533, 1988.

6. **Daniel, J.L., Molish, I.R., Robkin, L., and Holmsen, H.,** Nucleotide exchange between cytosolic ATP and F-actin-bound ADP may be a major energy-utilizing process in unstimulated platelets, *Eur. J. Biochem.,* 156, 677-684, 1986.

7. **Daniel, J.L., and Tuszynski, G.P.,** Platelet contractile proteins, in *Hemostasis and Thrombosis,* Colman, R.W., Hirsh, J.K., Marder, V.J., and Salzman, E.W., Eds., Lippincott, Philadelphia, 1987, 644-660.

8. **Eberle, M., Traynor-Kaplan, A.E., Sklar, L.A., and Norgauer, J.,** Is there a relationship between phosphatidylinositol triphosphate and F-actin polymerization in human neutrophils? *J. Biol. Chem.,* 265, 16725-16728, 1990.

9. **Holmsen, H.,** Mechanisms of platelet secretion, in *Platelets: Cellular Response Mechanisms and Their Biological Significance,* Rothman, A., Meyer, F.A., Gitler, C., and Silberberg, A., Eds., John Wiley and Sons, New York, 1980, 249-263.

10. **Holmsen, H.,** Phospholipids: The phosphatidylinositol cycle, the polyphosphoinositide cycle or phosphatidic acid? in *Platelet Responses and Metabolism,* Vol. III, Holmsen, H., Ed., CRC Press, Boca Raton, 1987, 121-135.

11. **Holmsen, H., Dangelmaier, C.A., and Holmsen, H.-K.,** Thrombin-induced platelet responses differ in requirement for receptor occupancy. Evidence for tight coupling of occupancy and compartmentalized phosphatidic acid formation, *J. Biol. Chem.,* 256, 9393-9396, 1981.

12. **Holmsen, H., Dangelmaier, C.A., and Rongved, S.,** Tight coupling of thrombin-induced acid hydrolase secretion and phosphatidate synthesis to receptor occupancy in human platelets, *Biochem. J.,* 222, 157-167, 1984.

13. **Kucera, G.L., and Rittenhouse, S.E.,** Human platelets form 3-phosphorylated phosphoinositides in response to alpha-thrombin U46619 or GTP gamma S, *J. Biol. Chem.,* 265, 5345-5348, 1990.

14. **Lages, B., Holmsen, H., Weiss, H.J., and Dangelmaier, C.A.,** Thrombin and ionophore A23187-induced dense granule secretion in storage pool-deficient platelets: Evidence for impaired nucleotide storage as the primary dense granule defect, *Blood,* 61, 154-162, 1983.

15. **Lapetina, E.G.,** Effect of pertussis toxin on the phosphodiesteratic cleavage of the polyphosphoinositides by guanosine 5'-0-thiotriphosphate and thrombin in permeabilized human platelets, *Biochim. Biophys. Acta,* 884, 219-224, 1986.

16. **Lassing, I., and Lindberg, U.,** Specificity of the interaction between phosphatidylinositol 4,5-biphosphate and the profilini actin complex, *J. Cell. Biochem.,* 37, 255-267, 1988.

17. **Majerus, P.W., and Cunningham, T.W.,** Pathway for the formation of D-3 phosphate containing phospholipids in PDGF stimulated NIH3T3 fibroblasts, *Biochem. Biophys. Res. Commun.,* 175, 568-576, 1990.

18. **Majerus, P.W., Ross, T.S., Cunningham, T.W., Caldwell, K.K., Jefferson, A.B., and Bansal, V.S.,** Recent insights in phosphatidylinositol signaling, *Cell,* 63, 459-65, 1990.

19. **Mauco, G., Dajeans, P., Chap, H., and Douste-Blazy, L.,** Subcellular localization of inositol lipids in blood platelets as deduced from the use of labeled precursors, *Biochem. J.,* 244, 757-761, 1987.

20. **Mauco, G., Dangelmaier, C.A., and Smith, J.B.,** Inositol lipids, phosphatidate and diacylglycerol share stearoylarachidonylglycerol as a common backbone in thrombin-stimulated human platelets, *Biochem. J.,* 224, 933-940, 1984.

21. **McKean, M.L.,** Phospholipid metabolism, in *Platelet Responses and Metabolism,* Vol. 2, Holmsen, H., Ed., CRC Press, Boca Raton, 1987, 279-286.

22. **Nishizuka, Y.,** The role of protein kinase C in cell surface signal transduction and tumour promotion, *Nature,* 308, 693-698, 1984.

23. **Rink, T.J., Sanches, A., and Hallam, T.,** Diacylglycerol and phorbol ester stimulate secretion without raising cytoplasmic free calcium in human platelets, *Nature,* 305, 317-319, 1983.

24. **Rink, T.J., Smith, S.W., and Tsien, R.Y.,** Cytoplasmic free Ca^{2+} thresholds and Ca-independent activation for shape change and secretion, *FEBS Lett.,* 148, 58-64, 1982.

25. **Serunian, L.A., Haber, M.T., Fukui, T., Kim, J.W., Rhee, S.G., Lowenstein, J.M., and Cantley, L.C.,** Polyphosphoinositides produced by phosphatidinositol 3-kinase are poor substrates for phospholipases C from rat liver and bovine brain, *J. Biol. Chem.,* 264, 17809-17815, 1989.

26. **Siffert, W., Fox, G., Muckenhoff, K., and Scheid, P.,** Thrombin stimulates Na^+/H^+ exchange across the human platelet plasma membrane, *FEBS Lett.,* 172, 272-274, 1984.

27. **Steen, V.M., and Holmsen, H.,** Current aspects of the (human) platelet responses, *Eur. J. Haematol.,* 38, 383-399, 1987.

28. **Steen, V.M., Tysnes, O.-B., and Holmsen, H.,** Synergism between thrombin and epinephrine in human platelets. Marked stimulation of the phosphoinositide metabolism, *Biochem. J.,* 253, 581-586, 1988.

29. **Tysnes, O.-B., Aarbakke, G.M., Verhoeven, A.J.M., and Holmsen, H.,** Thin-layer chromatography of polyphosphoinositides from platelet extracts, Interference by an unknown phospholipid, *Thromb. Res.,* 40, 329-338, 1985.

30. **Tysnes, O.-B., Verhoeven, A.J.M., Aarbakke, G.M., and Holmsen, H.,** Phosphoinositide metabolism in resting and thrombin-stimulated human platelets. Evidence against metabolic heterogeneity, *FEBS Lett.,* 218, 68-72, 1987.

31. **Tysnes, O.-B., Verhoeven, A.J.M., and Holmsen, H.,** Phosphate turnover of phosphatidylinositol in resting and thrombin-stimulated platelets, *Biochim. Biophys. Acta,* 889, 183-191, 1986.

32. **Tysnes, O.-B., Verhoeven, A.J.M., and Holmsen, H.,** Studies on the preferential incorporation of ^3H-glycerol over ^{32}P-orthophosphate into major phospholipids of human platelets, *Biochim. Biophys. Acta,* 930, 338-345, 1987.

33. **Tysnes, O.-B., Verhoeven, A.J.M., and Holmsen, H.,** Rates of production and consumption of phosphatidic acid upon thrombin stimulation of human platelets, *Eur. J. Biochem.,* 174, 75-79, 1988.

34. **Verhoeven, A.J.M., Mommersteg, M.E., and Akkerman, J.W.,** Quantification of energy consumption during thrombin-induced aggregation and secretion. Tight coupling between platelet responses and the increment in energy consumption, *Biochem. J.,* 221, 777-787, 1984.

35. **Verhoeven, A.J.M., Tysnes, O.-B., Aarbakke, G.M., Cook, C.A., and Holmsen, H.,** Turnover of the phosphomonoester groups of polyphosphoinositides in unstimulated human platelets, *Eur. J. Biochem.,* 166, 3-9, 1987.

36. **Verhoeven, A.J.M., Tysnes, O.-B., Horvli, O., Cook, C.A., and Holmsen, H.,** Stimulation of phosphate uptake in human platelets by thrombin and collagen. Changes in specific ^{32}P-labeling of metabolic ATP and polyphosphoinositides, *J. Biol. Chem.,* 262, 7047-7052, 1987.

37. **Vickers, J.D., and Mustard, J.F.,** The phosphoinositides exist in multiple metabolic pools in rabbit platelets, *Biochem. J.,* 238, 411-417, 1986.

38. **Vickers, J.D., and Mustard, J.F.,** Phospholipids: Fate of polyphosphoinositides, in *Platelet Responses and Metabolism,* Vol 2, Holmsen, H., Ed., CRC Press, Boca Raton, 1987, 137-150.

39. **Whitman, M., Downes, C.P., Keeler, M., Keller, T., and Cantley, L.C.,** Type I phosphatidylinositol kinase makes a novel inositol phospholipid, phosphatidylinositol-3 phosphate, *Nature,* 332, 644-646, 1988.

40. **Wilson, D.B., Neufeld, E.J., and Majerus, P.W.,** Phosphoinositide interconversion in thrombin-stimulated human platelets, *J. Biol. Chem.,* 260, 1046-1051, 1985.

Chapter 6

EPINEPHRINE-INDUCED PLATELET MEMBRANE MODULATION

Gundu H.R. Rao and James G. White

INTRODUCTION

Studies from our laboratory have demonstrated the existence of an intrinsic mechanism capable of restoring the sensitivity of refractory blood platelets to the action of agonists.[89,90,93,94,95,96,99] This newly discovered, novel mechanism, termed "membrane modulation,"[95] is intrinsic to platelet plasma membranes and is capable of securing irreversible aggregation without the mediation of newly generated cyclic endoperoxide, thromboxane A_2, platelet activating factor (PAF) or dense granule contents such as calcium, serotonin (5-HT) and adenosine diphosphate (ADP).[89,90,98,100] It is mediated by alpha-adrenergic receptors, is calcium dependent, and facilitates calcium uptake, fibrinogen binding and phosphorylation of some proteins.[21,73,74,96] Epinephrine-induced membrane modulation and restoration of function in refractory platelets seems to be independent of enhanced phosphoinositol metabolism, elevation of cytosolic calcium and phosphorylation of the myosin light chain.[88] It is a novel, independent salvage mechanism capable of restoring sensitivity of the plasma membrane to the action of agonists in platelets with compromised function. This chapter reviews findings related to epinephrine-induced facilitation of platelet function with special emphasis on the mechanism of membrane modulation.

FACILITATING INFLUENCE OF EPINEPHRINE ON PLATELET FUNCTION

O'Brien, in his classic paper on the influence of adrenaline (epinephrine) and anti-epinephrine compounds on blood platelets, described the ability of epinephrine to induce stickiness and aggregation. He also demonstrated that antagonists, such as phentolamine, prolonged bleeding time. Based on his own observations and earlier reports, he suggested that epinephrine or norepinephrine may play a role in capillary hemostasis.[71] Administration of nicergoline, an alpha-adrenoreceptor antagonist, seems to prevent human platelet aggregation.[82] Platelet-mediated formation of thrombi in stenosed canine arteries has been shown to be inhibited by nicergoline,[6] suggesting a role for alpha-adrenoceptor modulation in platelet activation and thrombus formation. Human platelets stirred with epinephrine (1-10 μM) aggregate and release the contents of their granules. It is believed that this agent causes platelets to aggregate without changing their shapes. However, Milton and Frojmovic[67] demonstrated the importance of pseudopods in epinephrine-induced aggregation. Although human platelets anticoagulated with citrate aggregate in response to epinephrine, washed platelets resuspended in various buffers do not. Epinephrine can, however, potentiate the action of most agonists in citrated and in washed platelet suspensions. Similarly, in platelets of rabbits, cats, dogs, mice and in thrombocytes of turkeys, epinephrine potentiates the action of other agonists. In platelets of cows, horses, sheep, guinea pigs and pigs, epinephrine exerts little effect. Guinea pig platelets seem to lack alpha$_2$-adrenoceptors.

Epinephrine-induced potentiation of the action of other agonists is well documented.[11,12,20,23,27,37,43,51,117,125,128] Epinephrine not only potentiates the action of physiological agonists such as ADP, collagen, and thrombin, but also the action of phorbol ester, the calcium ionophore, A23187, PAF and bacteria/platelet interaction. It has been shown to significantly shorten the lag phase response of platelets to the action of collagen and bacteria (Figures 1 and 2). The exact mechanism by which epinephrine influences the formation of pseudopods, the development of stickiness, and potentiates the action of other agonists, is not yet clear. Some of the suggested mechanisms are discussed later in this chapter.

Facilitating Influence of Epinephrine on Collagen Induced Platelet Aggregation

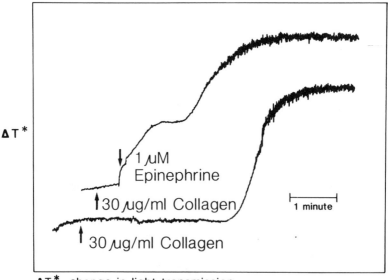

ΔT^* - change in light transmission

FIGURE 1. Collagen-induced stimulation of human platelets was accompanied by a prolonged lag phase followed by irreversible aggregation. Epinephrine at subthreshold concentration significantly shortened the lag phase and potentiated collagen-induced aggregation.

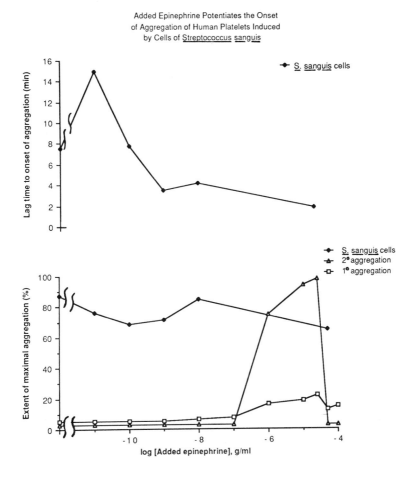

Added Epinephrine Potentiates the Onset
of Aggregation of Human Platelets Induced
by Cells of Streptococcus sanguis

FIGURE 2. *Streptococcal sanguis*-induced stimulation of human platelets is accompanied by a prolonged lag phase followed by irreversible aggregation and secretion of dense body contents. Epinephrine at concentrations lower than 1 µM does not cause human platelet aggregation. However, at nanomolar concentrations, epinephrine shortens the lag phase and potentiates *S. sanguis*-induced platelet aggregation (Courtesy of G. MacFarlane, L. Krishhan and M. Herzberg.)

ALPHA ADRENOCEPTORS AND PLATELET FUNCTION

Hormones such as epinephrine and norepinephrine act through a specific receptor on the plasma membrane. The receptors are coupled to specific effector molecules via GTP binding proteins.[57] Extensive pharmacological studies have identified several distinct types of adrenergic receptors on blood platelets.[39,64] Those that are well characterized include $alpha_1$, $alpha_2$, $beta_1$, and $beta_2$. Epinephrine has been shown to exert its effect in platelets via alpha-adrenergic receptors. The number of alpha-adrenoceptor sites per platelet is 200-400. All of the subtypes of adrenergic receptors are activated by epinephrine.[53] The gene for the human platelet alpha$_2$-adrenoceptor has been cloned and the partial amino acid sequence of the purified receptor has been analyzed.[54] A schematic orientation of the human alpha$_2$-adrenoceptor with respect to plasma membrane is shown in Figure 3. Like other receptors described to date, alpha$_2$-adrenoceptor has a short extracellular amino terminal, the usual hydrophobic seven transmembrane domains, cytoplasmic loops and the intracellular carboxyterminal. Of the various domains, the region of greatest homology seems to be amino acid residues between 111-136 (third transmembrane and second cytoplasmic loop). Unlike other receptors, the carboxyterminal end of the alpha$_2$-adrenoceptor lacks serine or threonine residues. The third cytoplasmic loop contains 19 serine and threonine residues, potential sites for phosphorylation. Although epinephrine is thought to activate platelets via this receptor, little is known about the molecular mechanisms involved in receptor-mediated signal transduction.[53,54,57,79,80]

Human Platelet Alpha-2 Adrenergic Receptor

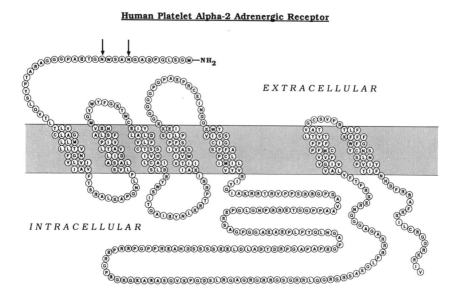

FIGURE 3. Schematic orientation of the structure of human platelet at alpha$_2$-adrenoceptors as it may be organized within the membrane. It is similar in structure to other receptors such as beta$_1$ and beta$_2$ adrenoceptors. Of the many amino acids of the membrane spanning domain, 45% are homologous with the beta$_1$ adrenoceptor. There seems to be considerably less homology between this receptor and others in intracellular and extracellular domains. A major difference between this receptor and others seems to be the lack of phosphorylation sites (serine and threonine residues) in the carboxy terminal. However, the cytoplasmic loops contain significant phosphorylation sites, especially the third cytoplasmic loop (total of 19 serine and threonine residues) (Courtesy R.J. Lefkowitz and J.W. Regan).

Investigations of alpha$_2$-adrenoceptor density in human platelets, under a variety of conditions, have resulted in inconsistent findings.[81] There are conflicting reports regarding the number of receptor sites and the functional state of these receptors. Other factors that influence receptor sites include drug treatment, obesity, physiology, and various disease states.[19,26,34,58,59,72,106,118, 120,121,122,130,131] Attempts to correlate the density of receptor sites to the functional responses of platelets have not yielded conclusive results.[34,40,65,81,122] Our laboratory has shown that phospholipase A$_2$ (LA) treatment of platelets results in loss of some receptor sites and induces a refractory state to the action of epinephrine.[103] Although these platelets do not aggregate when stirred with the catecholamine, the potentiating influence of epinephrine could still be demonstrated. If LA-treated platelets are made dysfunctional by drug exposure (aspirin/prostaglandin E1 (PGE1)), sensitivity to the action of other agonists can be restored by epinephrine treatment. Results of these studies suggest that loss of some receptor density may not lead to total dysfunction. There are many discrepancies in the published literature concerning the receptor density, functional state and other properties of the platelet alpha adrenoceptor. It is possible that these receptors exist in different states with varying sensitivity to pharmacological agents. Further pharmacological studies are essential to develop the consistent data required to establish a correlation between the density of these receptors and the functional response they elicit.

DISCOVERY OF PLATELET MEMBRANE MODULATION

Early studies on platelet biochemistry and physiology favored a concept that considered ADP the final mediator of all platelet aggregation.[7,33,38,48,107] The discovery of prostaglandins and intensive research in this area added further strength to this concept.[62,116] It is well documented that the cyclooxygenase inhibitors block the second wave response of platelets to physiological agonists. This is also true of ADP scavengers.[25,32,76,129] Since ADP and thromboxane are capable of inducing aggregation independent of other agents, the two mechanisms are considered essential for irreversible platelet aggregation.

A major breakthrough in our efforts was the discovery of a mechanism in dog platelets which secured irreversible aggregation independent of prostaglandin endoperoxide/thromboxane involvement. About two-thirds of mongrel dogs have platelets which do not aggregate when stirred with sodium arachidonate (AA).[44] However, the AA-refractory dog platelets undergo shape change and produce normal quantities of thromboxane when stirred with AA. Further studies revealed that small amounts of epinephrine, which do not cause aggregation of dog platelets, could restore the sensitivity of these cells to the action of AA.[45,46,47]

In view of this observation, platelets of normal donors were tested to see if such a heterogeniety could be demonstrated. A single donor was found to have platelets that did not respond to AA stimulation. However, extensive studies demonstrated that the enzymes in platelets from this individual lacked the ability to convert arachidonic acid to cyclooxygenase products.[99] Despite the biochemical lesion, platelets from this individual aggregated irreversibly when exposed to epinephrine and AA. Studies from our laboratory also demonstrated that exposure of normal platelets to small quantities of adenylate cyclase stimulators, such as prostacyclin (PGI_2), PGE_1, prostaglandin D_2 (PGD_2) and forskolin, made them specifically refractive to AA stimulation. However, similar to dog platelets and cyclooxygenase deficient platelets, epinephrine exposure restored the sensitivity of these platelets to the action of AA. Elevation of intracellular cyclic adenosine monophosphate (cAMP) levels could not be demonstrated in any of these studies; therefore, it was assumed that epinephrine must be working through an entirely different mechanism. This newly discovered mechanism was termed "platelet membrane modulation." [95]

IRREVERSIBLE AGGREGATION INDEPENDENT OF THE MEDIATION BY CYCLOOXYGENASE PRODUCTS OR DENSE GRANULE CONTENTS

The fact that cyclooxygenase-deficient platelets could be made to respond to the action of AA by epinephrine prompted further studies on aspirin-exposed platelets. Platelets from normal individuals 16-24 hours after ingesting 360 mg aspirin were unresponsive to AA and gave single waves in response to threshold concentrations of thrombin, ADP and epinephrine. However, prior exposure of the aspirin-treated cells to concentrations of epinephrine too low to cause aggregation restored the sensitivity of cells to all agents,[90,96] including AA. Restoration of the ability of aspirin-treated cells and cyclooxygenase-deficient platelets to undergo irreversible aggregation after treatment with a low dose of epinephrine did not improve in the ability of cells either to synthesize prostaglandin endoperoxides or to secrete granule contents.[89] Studies by Cameron and Ardlie[12] concur with our observations regarding the ability of epinephrine to influence the aggregation response of aspirin-treated platelets to AA. However, they also showed, contrary to other reports, the ability of epinephrine to stimulate the cyclooxygenase pathway under the inhibitory influence of acetylsalicylic acid. Studies from other laboratories have demonstrated that epinephrine potentiation of PAF-induced aggregation of aspirin-treated platelets is independent of cyclooxygenase metabolites of AA.[31,128]

To establish that secretion of granule contents was not involved in the irreversible aggregation mediated by the new mechanism of membrane modulation, it seemed worthwhile to evaluate this mechanism in storage pool deficient platelets. The patients chosen for these studies were well characterized regarding the specific defect, Hermansky-Pudlak syndrome (HPS), and had no dense bodies in their platelets. Eighteen hours after ingestion of aspirin, cyclooxygenase activity was completely blocked in their cells. However, when a low dose of epinephrine was used in combination with other agonists, the platelets from patients with HPS aggregated irreversibly without any measurable secretion of adenine nucleotides.[89] It is reasonable to conclude that the intrinsic mechanism of membrane modulation is capable of promoting irreversible platelet aggregation in the absence of secretion and prostaglandin synthesis.

In addition to the aggregation studies described in previous sections, the mechanism of membrane modulation was explored during platelet-vessel wall interaction (Baumgartner technique) using cells from normal donors and patients with storage pool disorders (SPD). Untreated control platelets formed normal thrombi on the exposed rabbit subendothelium when perfused through a Baumgartner chamber. Aspirin treatment of platelets significantly reduced the frequency of thrombi. Addition of epinephrine to aspirin-treated normal platelets or cells from HPS patients reversed the inhibitory effect of the drug.[85,86]

The micropipette aspiration technique was utilized to demonstrate membrane modulation at the single cell level. Pipette tips had an internal diameter of 0.7 μm or less, and the influence of aspirin, indomethacin and ibuprofen on the human platelet resistance to aspiration into micropipettes was evaluated. Aspirin increased the length of platelet extensions into the micropipette, indicating that membrane fluidity had increased. Other cyclooxygenase inhibitors such as ibuprofen and indomethacin did not increase platelet deformability. Pretreatment of platelets with epinephrine prevented aspirin from altering platelet pliability, and exposure of platelets to epinephrine after aspirin treatment reversed the increased pliability induced by the drug.

Clonidine effectively blocked the influence of epinephrine on aspirin platelets, suggesting that epinephrine-induced alterations were alpha-adrenoceptor mediated.[10,115] Results clearly demonstrated the intrinsic mechanism of membrane modulation at a single cell level. Folts *et al.*[28,29,30] demonstrated that acute platelet thrombi form periodically in stenosed dog, pig coronary arteries and in monkey and rabbit carotid arteries. Thrombus formation could be abolished by 5 mg/kg of aspirin administered intravenously. In 75% of the animals tested, infusion of epinephrine (0.4 μg/kg/min) was shown to cause renewal of thrombus formation.

MECHANISM INVOLVED IN EPINEPHRINE-INDUCED RESTORATION OF FUNCTION IN ASPIRIN-TREATED PLATELETS

Chignard and associates[14,15] demonstrated that PAF was released from platelets activated with the calcium ionophore, A23187, thrombin or collagen. They also suggested that this novel agonist could induce aggregation independent of AA metabolites or released ADP.[8] Therefore, the possibility that epinephrine-induced restoration of function in aspirin-treated platelets could be due to the action of newly generated PAF had to be examined. Aspirin-treated platelets were made refractory to the action of PAF by prior exposure to low concentrations of the alkenylacetyl glycerophosphocholine. Platelets exposed to this analogue of PAF did not aggregate when tested with concentrations as high as 4 μM PAF. However, PAF-refractory, aspirin-treated platelets aggregated irreversibly when stimulated by epinephrine and AA.[98,100] These results indicated that epinephrine-induced restoration of function in refractory platelets was not mediated by PAF.

Aspirin treatment inactivates platelet cyclooxygenase and prevents conversion of arachidonic acid into prostaglandin endoperoxides. However, aspirin does not prevent release of arachidonic acid from membrane phospholipids in stimulated platelets or its conversion by lipoxygenase into hydroperoxy and hydroxy fatty acids. Therefore, it was critical to determine whether AA metabolites generated via the lipoxygenase pathway mediate the restoration in function achieved by membrane modulation. To ascertain whether inhibition of lipoxygenase enzymes prevents membrane modulation, two specific drugs were used: U53119 (4, 7, 10, 13-eicosatetraynoic acid) and ETYA (5, 8, 11, 14-eicosatetraynoic acid). ETYA totally blocked conversion of arachidonic acid by both the pathways, whereas U53119 specifically inhibited the lipoxygenase pathway. The inhibitory concentrations of these fatty acids failed to prevent AA-induced irreversible aggregation of cyclooxygenase-inhibited platelets in the presence of epinephrine. Results of these studies demonstrated that the generation of lipoxygenase metabolites of AA are not essential for securing irreversible aggregation of aspirin-treated cells by the mechanism of membrane modulation.[92,102]

STUDIES ON MEMBRANE MODULATION IN VARIOUS PLATELET DISORDERS

Studies from our laboratory have demonstrated that platelets from patients with SPD or cyclooxygenase deficiency can be made to respond in a normal fashion to the action of aggregating agents by membrane modulation. In view of these observations, we routinely test platelets of patients with various disorders to see if the presence of the intrinsic mechanism can be demonstrated. We have studied membrane modulation in platelets of patients with the grey platelet syndrome, the Bernard-Soulier syndrome, Glanzmann's thrombasthenia, May-Hegglin anomaly, HPS, Chediak-Higashi syndrome, Wiskott-Aldrich syndrome, Ehlos-Danlos syndrome, cyclooxygenase deficiency, afibrinogenemia and various leukemias. The ability of epinephrine to potentiate the action of agonists was demonstrated in all of the disorders tested, except Glanzmann's thrombasthenia.[86] These results suggest that the intrinsic mechanism of membrane modulation may maintain platelet function in a normal fashion in microcirculation and, thereby, prevent the manifestation of severe bleeding in these patients.

EPINEPHRINE-INDUCED RESTORATION OF FUNCTION IN DRUG-INDUCED REFRACTORY PLATELETS

Studies of animal (turkey, dog, rabbit, pig and monkey) and human control platelets, as well as drug-induced refractory platelets, have shown that irreversible aggregation could be achieved independent of prostaglandin synthesis, or the release reaction.[86] To evaluate the platelet activation, inactivation and reactivation phases by pharmacological manipulation, a simple assay system was designed using the classical aggregometry.[101] First, platelets were stimulated with various agonists. After the aggregation response reached a plateau, the effects of anti-platelet drugs were tested. If a drug was potent, aggregate disaggregation occurred. The disaggregated cells could be challenged for a second time with agonists. Indeed, these cycles could be repeated more than twice with proper manipulation.[101] Studies from our laboratory demonstrated that agents that elevated cAMP, scavengers of ADP, phospholipase inhibitors, some calcium antagonists and membrane-active compounds, caused disaggregation of platelet aggregates.[93,94] These studies show that platelets retain their intrinsic ability to respond to the action of agonists even after they have undergone aggregation and release reaction.

Using this well-characterized model, fibrinogen binding and protein phosphorylation were followed during aggregation-disaggregation-reaggregation cycles.[21] During aggregation induced by ADP, fibrinogen binding and extensive phosphorylation of some of the proteins could be demonstrated. Stimulation of adenylate cyclase with PGI_2 or PGE_1 and a rise in intracellular cAMP levels induced rapid dissociation of bound fibrinogen and initiated dephosphorylation of proteins. Addition of epinephrine to these cells lowered the level of cAMP. Further challenge with AA caused reaggregation. This reactivation of platelets was accompanied by reassociation of fibrinogen and rephosphorylation of proteins.[21] Peerschke[78] demonstrated that epinephrine induced maximal aggregation of aspirin-treated platelets stimulated with thrombin or ADP by significantly enhancing fibrinogen receptor exposure, independent of the cyclooxygenase-mediated release reaction.

Thus, epinephrine exposure of human platelets made refractory by the action of cyclooxygenase inhibitors (aspirin, indomethacin, ibuprofen), adenylate cyclase stimulators PGE1, PGI_2, PGD_2, forskolin, adenosine, membrane-active drugs (chlorpromazine, quinicrine, trifluoperazine), calcium antagonists (quinoline derivatives), antibiotics (carbenicillin, penicillin), receptor antagonists (13-Azaprostanoic acids, SKF 96148, SKF 95587) restored the sensitivity to the action of other agonists. In addition, our studies have shown that by careful manipulation, the membrane response to the agonists can be obtained more than once.[86,95,96,97,101]

IS MEMBRANE MODULATION DIFFERENT FROM SYNERGISM?

Since two agonists are needed to elicit the irreversible aggregation response of refractory platelets, the possibility of synergism had to be examined. Two types of experimental conditions were used to demonstrate that membrane modulation is an independent phenomenon. First, the ability of epinephrine to induce irreversible aggregation of aspirin-treated human platelets to the action of thromboxane A_2 was tested (Figure 4). Epinephrine at nanomolar concentrations (50-100 nM) potentiated the action of thromboxane, whereas other agonists such as ADP, collagen, thrombin or A23187 had no such effect at very low concentrations.[46,47,86] Similar results have been obtained in studies with *S. Sanguis*-induced platelet activation (Figure 2) (M.C. Herzberg, unpublished data).

In a second series of experiments, prostaglandin-dissociated (PGI_2/PGE_1), ADP aggregates were used to test our hypothesis. PGI_2-dissociated platelets were challenged for their response to various agonists. The dissociated cells were refractory, but following exposure to epinephrine they regained their sensitivity to all agonists (Figure 5). Yet, when challenged with a combination of thrombin and AA, ADP and AA, or collagen and AA, PGE_1-dissociated cells did not aggregate irreversibly (Figure 6). These studies, done under two different conditions, demonstrated that refractory platelets cannot be stimulated by just any two agonists, and they laid to rest the suggestion that membrane modulation is a manifestation of synergism.

MECHANISMS INVOLVED IN EPINEPHRINE-INDUCED FACILITATION OF PLATELET FUNCTION

Although epinephrine-induced potentiation of various platelet antagonists is well documented, the exact mechanism by which this hormone facilitates the action of other agonists remains unknown.[3,11,12,43,50,56,66,77,117,119,124] Many attempts have been made to denote one or another physical, mechanical or biochemical event associated with platelet activation as the key mechanism by which epinephrine potentiates the action of other agonists.[11,17,18,21,73,80] The presence of extracellular calcium, fibrinogen and cell-to-cell contact (stirring) are essential features for promoting platelet stickiness and aggregation. Therefore, earlier studies focused on epinephrine's ability to modulate calcium uptake or levels, as well as its influence on induction of binding sites for fibrinogen or promotion of fibrinogen binding.[111,112]

Studies using chlortetracycline (CTC) as the calcium fluorophore suggested that epinephrine promoted calcium uptake by the plasma membrane in aspirin-treated platelets.[73,74] Using radiolabeled calcium, Brass and Shattil[9] could find no evidence for epinephrine's ability to increase net influx of Ca^{2++} into platelets. However, they demonstrated that epinephrine stimulation enhanced

Epinephrine (EPI) Potentiation of Thromboxane A₂ Stimulated Human Platelet Aggregation

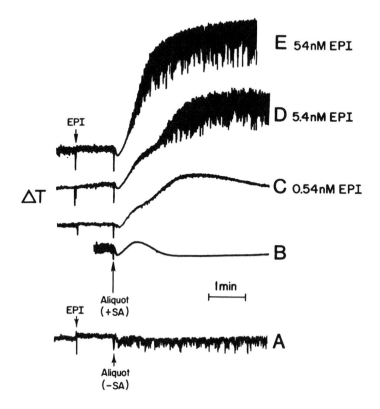

FIGURE 4. Low concentrations of epinephrine and arachidonate (AA) used in these studies did not cause aggregation of platelets (A). Small quantities of AA (μM) were incubated with prostaglandin E_1 (PGE$_1$)-treated, washed platelets for 30 seconds to generate thromboxane A$_2$; PGE$_1$ was used in these studies to prevent platelet stimulation and secretion by AA. Aliquots of a sample in which PGE$_1$-treated, washed platelets were exposed to a known quantity (μM) of AA, did not induce platelet aggregation (B). Epinephrine at μM concentrations did not cause platelet aggregation (C). However, epinephrine-exposed platelets, upon addition of aliquots from AA-incubated platelets, caused irreversible aggregation (E and D).

surface membrane-bound calcium and suggested that this pool may participate in several membrane reactions involved in platelet activation. Several studies have failed to demonstrate epinephrine-induced cytosolic calcium mobilization in the absence of cyclooxygenase activity.[9,16,22,91] Clare and Scrutton[16] also failed to show epinephrine-induced calcium influx into platelets; they had raised the question about the specificity of verapamil as a calcium channel blocker in earlier investigations.

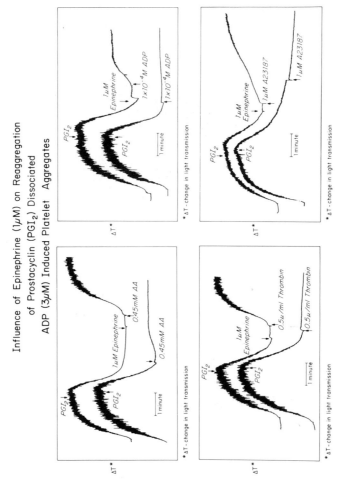

FIGURE 5. Addition of prostacyclin (PGI$_2$) following ADP-induced aggregation caused dissociation of aggregates. Dissociated cells were refractory to the action of all agonists. However, these refractory cells, when exposed to epinephrine, regained the sensitivity to the action of agonists and underwent irreversible aggregation. (From Rao, G. H. R., Reddy, K., and White, J., *Prostaglandins Med.,* 4, 385, 1980. With permission.)

Studies from our laboratory could not demonstrate the ability of verapamil to block calcium influx caused by agonists such as thrombin and endoperoxide mimetic (U46619).[86] Its ability to inhibit PAF-induced calcium flux may be related to its known effect on the alpha$_2$-adrenoceptor. Verapamil has been shown to antagonize epinephrine binding. Similarly, it may prevent PAF binding to platelets as well. Several studies support or postulate that epinephrine, acting via the alpha$_2$-adrenoceptor, modulates phospholipase C (PLC), and enhances phosphastidyl inositol (PI) turnover or phosphatidyl 4,5-biphosphate

Response of Prostaglandin E₁, (PGE₁) Dissociated ADP Aggregates
to the Action of Two Agonists

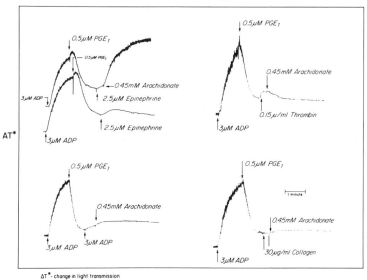

ΔT*- change in light transmission

FIGURE 6. Addition of PGE_1 following ADP-induced aggregation caused dissociation of aggregates. Dissociated cells were not refractory to the action of single agonists or to combinations of ADP, thrombin, or collagen with AA. Only epinephrine was able to potentiate the action of PGE_1-dissociated cells to the action of other agonists. (From Rao, G. H. R. and White, J., *Agents and Actions*, 16, 425, 1985. With permission.)

(PIP_2) hydrolysis.[2,11,55,83,125] However, the exact mechanism by which epinephrine facilitated these events is not clear.[109]

In rabbit platelets, epinephrine-induced potentiation of ADP aggregation seems to facilitate mechanisms by which ADP induces turnover of phosphatidylinositol 4-phosphate (PIP) and hydrolysis of PI. Therefore, epinephrine, which has little effect on PI metabolism, seems to influence the ability of ADP to modulate the PI pathway. In human platelets, subthreshold concentrations of epinephrine and ADP together induce release of granule contents; when used singly, they are ineffective. Reaction velocities induced by the combination seem to be slower than those induced by thrombin. Indomethacin significantly blocks the release reaction induced by epinephrine and ADP, whereas it exerts no effect on thrombin. Based on these results, Kawahara *et al.*[49] concluded that thromboxane A_2 may be involved in 5-HT secretion and in a 40 kDa protein phosphorylation caused by epinephrine and ADP. Similarly, vasopressin and epinephrine used singly do not significantly elevate cytosolic calcium levels.[11] However, in combination, these agents caused significant elevation of cytosolic calcium in Quin 2-loaded platelets. In these studies, epinephrine was shown to potentiate the formation of phosphatidic acid, suggesting its influence on a PLC-dependent mechanism. Based on these findings, Bushfield *et al.*[11] concluded that epinephrine, either by direct alpha$_2$-adrenoceptor occupancy or as a consequence of suppression of cAMP levels, modulates the G-protein-mediated hydrolysis of PIP$_2$.

In addition to the calcium that can be mobilized from internal membrane pools, agonists can induce influx of this cation from the exterior medium. This influx can be followed by using Mn^{2+} as a chase ion in Quin 2-loaded or Fura 2-loaded platelets. Epinephrine has been shown to enhance the Ca^{2+} influx induced by U46619, thrombin and ADP. Thompson *et al.*[125] suggested that epinephrine, acting via the alpha$_2$-adrenoceptor, modulates receptor-PLC coupling. They concluded that such modulation was not mediated by inhibition of adenylate cyclase. Although the exact mechanism by which epinephrine modulates stimulus-activation-coupling is not clear, the majority of these studies postulate that the action of epinephrine is due to its known inhibitory effect on adenylate cyclase mediated by the inhibitory GTP-binding protein, G$_i$.

Crouch and Lapetina[22] demonstrated the ability of epinephrine to potentiate the action of thrombin in a variety of experimental conditions. They showed that prolonged incubation of platelets with thrombin or exposure to PGI2 desensitized both PLC and calcium mobilization by thrombin. Using these refractory models, they demonstrated the ability of epinephrine to resensitize this mechanism via alpha$_2$-adrenoceptor stimulation. Based on their findings, they have suggested a scheme for platelet activation-receptor desensitization and epinephrine-induced resensitization (Figure 7). According to this scheme, the G$_i$ protein common to thrombin and alpha$_2$-adrenoceptor maintains the receptor coupling to PLC. When this protein is phosphorylated by protein kinase C (PKC), the action is inhibited. Epinephrine, via alpha$_2$-adrenoceptor modulation by an as yet unknown mechanism, causes the release of new alpha$_i$ subunits that couple with PLC and induce resensitization of this receptor to further challenges of thrombin.

Banga and associates[2] probed the dual role of glycoproteins IIb/IIIa and epinephrine (Figure 8) using a variety of inhibitors such as yohimbine, aspirin, indomethacin, ONO-RS-082, and an antibody for glycoprotein IIb-IIIa. Based on the results of their studies, they proposed: "...that epinephrine, in promoting exposure of glycoprotein IIb/IIIa sites for fibrinogen binding, leads to a cytoplasmic alkalinization, which in conjunction with local shifts in Ca^{2+}, promotes low-level activation of phospholipase A. The resulting free arachidonic is converted to cyclooxygenase products, which, potentiated by epinephrine, activate phospholipase C. This further amplifies the initial stimulatory response."

Epinephrine-mediated alpha$_2$-adrenoceptor modulation results in the inhibition of adenylate cyclase.[60] However, this effect of epinephrine will not lower the basal levels of intracellular cAMP. Since sodium has been shown to modulate inhibition of adenylate cyclase in broken cell preparations, there is consid-

1. PLATELET ACTIVATION AND RECEPTOR DESENSITIZATION

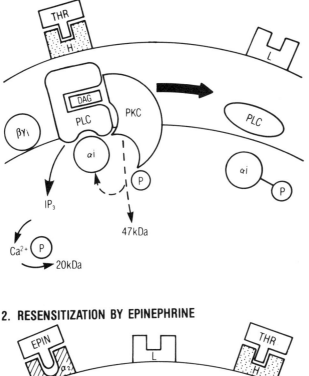

2. RESENSITIZATION BY EPINEPHRINE

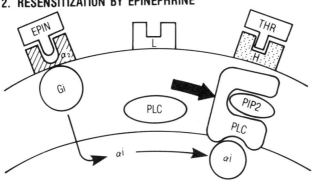

FIGURE 7. Model for the action of thrombin (THR) and epinephrine (EPIN) on human platelets. 1) Thrombin receptors can exist either coupled (H) or uncoupled (L) to phospholipase C (PLC), depending on whether alpha$_i$ (αi) is associated with the complex. In the agonist-occupied state, the thrombin receptor-PLC complex is able to accept and hydrolyze the phosphatidylinositol 4,5-bisphosphate (PIP2) substrate. Activation of PLC by thrombin induces inositol triphosphate (IP$_3$) production which mobilizes intracellular Ca^{2+} stores and participates in 20 kDa protein phosphorylation. The produced diacylglycerol (DAG) induces protein kinase C (PKC) to translocate to the plasma membrane near the receptor. PKC induces cellular responses in conjunction with Ca^{2+}, such as through 47 kDa protein phosphorylation, and PKC also phosphorylates αi. The phosphorylated αi is inactive and cannot maintain the thrombin receptor coupled to PLC, which results in desensitization of the thrombin receptor. 2) Stimulation of the α_2-adrenergic receptor with epinephrine (EPIN) releases more αi to the cytosol and re-couples the thrombin receptor to PLC (Courtesy E.G. Lapetina).

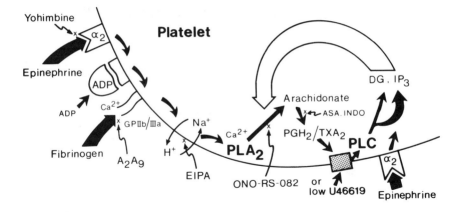

FIGURE 8. Dual Function of Epinephrine in Platelet Activation. The scheme shows the current knowledge of the mechanism of epinephrine action based upon data reported by Banga *et al.*, (Proceedings of the National Academy of Sciences 83, 9197, 1988) and in the literature. Jagged arrows indicate receptor antagonists and inhibitors: thromboxane A_2, TXA_2; prostaglandin H_2, PGH_2; aspirin, ASA; indomethacin, INDO; endoperoxide mimetic, U46619; glycoprotein IIb/IIIa antibody, A_2A_9; diacylglycerol, DG; inositol triphosphate, IP_3; phospholipase A_2, PLA_2; phospholipase C, PLC. Based on the results obtained by the use of a variety of antagonists and inhibitors, it is postulated that the functioning of the α_2-receptor in platelets is a dual one, as an initiation of events leading to TxA_2 formation and as potentiation of TxA_2-caused events (Courtesy S.E. Rittenhouse).

erable interest in the role of Na^+ in epinephrine-induced platelet activation.[61] It has been shown that removal of extracellular sodium results in diminished rate and extent of platelet aggregation and secretion in response to epinephrine.[17,18] Furthermore, it is postulated that platelet $alpha_2$-adrenoceptor may be a Na^+/H^+ antiport. A well-recognized event that follows stimulus-provoked activation of platelets is Na^+/H^+ exchange and a transient alkalinization of the platelet interior.[123] This alkalinization appears to promote LA activation at the concentration of Ca^{2+} available following cell activation. The generated AA metabolites seem to induce both secondary activation and the PI turnover essential for contraction-secretion coupling.[60] Using fluorescent dye 2',7'-bis(carboxyethyl) 5,6-(carboxyfluorescein), Zavoico *et al.*[134] followed intracellular pH changes under a variety of experimental conditions with thrombin and calcium ionophores A23187 and ionomycin as agonists. Based on their observations, they concluded that mobilization of cytosolic calcium enhances H^+ production, and subsequent elevation in pH is not related for receptor occupancy or increase in calcium levels. They also concluded that Na^+/H^+ exchange induced by thrombin may or may not be related to regulation of H^+ or PKC. Contrary to these observations, it has been shown that both activation of PKC, as well as PLC-induced Na^+/H^+ exchange.[114] This activation is inhibited by agents that elevate cAMP or inhibit PKC. Although Na^+/H^+ exchange has been shown to occur as a result of agonist-induced platelet activation, it is not clear what specific mechanism links this event with other events following cell activation. Siffert and Akkerman[113] suggested that the C kinase exerts its effect via Na^+/H^+ exchange and facilitates alkalinization and calcium mobilization. However, Banga et al.,[2] using an antibody (A_2A_9) to glycoprotein IIb/IIIa, demonstrated that alkalinization of cytoplasmic pH was a function of fibrinogen interaction with the GPIIb/IIIa complex, whereas Zavoico and Cragoe[133] have shown that calcium mobilization can occur independent of acceleration of Na^+/H^+ exchange in thrombin-stimulated platelets.

These studies suggest that extracellular Na^+ promotes ligand binding and potentiates or amplifies the agonist-induced signal. Epinephrine, via $alpha_2$-adrenoceptor modulation, promotes Na^+/H^+ exchange and increases cytoplasmic pH. Subsequent activation of PKC and fibrinogen binding facilitates calcium mobilization. Epinephrine appears to activate a low level of LA activity and indirectly potentiates the other PLC pathway.[2] Although data are not conclusive, epinephrine seems to potentiate all the earlier events associated with platelet activation as well as actions of the newly generated biologically active molecules and released dense body contents.

MECHANISMS INVOLVED IN EPINEPHRINE-INDUCED MEMBRANE MODULATION IN DRUG-INDUCED REFRACTORY PLATELETS

Studies in our laboratory focused on epinephrine-induced membrane modulation in three drug-induced refractory models: PGE_1/PGI_2; aspirin/ibuprofen; Quin 2/CTC, exposed cells.[104] The specific purpose of these studies was to determine whether the alterations in adrenoceptor numbers, changes in Na^+/H^+ exchange leading to alkalinization of the cytosol, enhanced fibrinogen binding or accelerated phosphoinositide hydrolysis (PLC-dependent/PKC-dependent pathways) underlie the epinephrine-induced restoration of membrane sensitivity to the action of physiological agonists (Figure 9).

Based on structure-activity-relationship, Rossi et al.[105] reported that the order of potency for catecholamines in inducing platelet aggregation was epinephrine > norepinephrine > epinine > dopamine. Epinephrine undergoes oxidation. To determine if epinephrine oxidation reduces the biological activity, a simple assay was devised. A potentiostat (Bioanalytical Instruments) was used to generate a constant 6-volt electrical field, and epinephrine was subjected to its action. Using aliquots from Stock solution subjected to a 6-volt current, it was possible to show a correlation between the degree of oxidation and a corresponding loss in epinephrine's ability to activate platelets (Figure 10). Platelets exposed to oxidized products of epinephrine were not sensitive to a further epinephrine challenge.[103]

These results suggest that epinephrine binding is essential for alpha-adrenoceptor stimulation and that oxidation of the drug impairs its ability to stimulate platelets, although oxidized products may also react with the receptor and

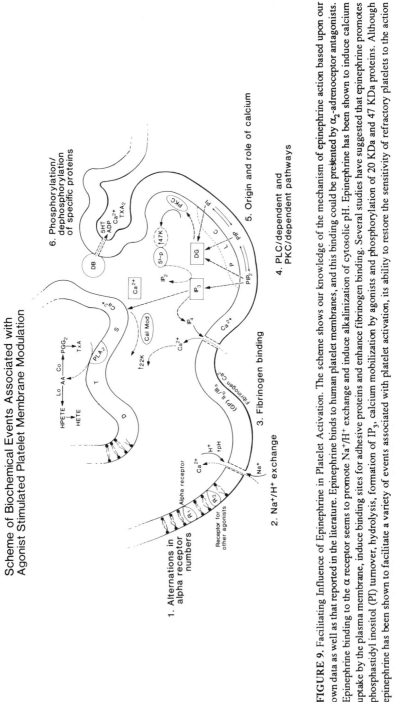

FIGURE 9. Facilitating Influence of Epinephrine in Platelet Activation. The scheme shows our knowledge of the mechanism of epinephrine action based upon our own data as well as that reported in the literature. Epinephrine binds to human platelet membranes, and this binding could be prevented by α_2-adrenoceptor antagonists. Epinephrine binding to the α receptor seems to promote Na^+/H^+ exchange and induce alkalinization of cytosolic pH. Epinephrine has been shown to induce calcium uptake by the plasma membrane, induce binding sites for adhesive proteins and enhance fibrinogen binding. Several studies have suggested that epinephrine promotes phosphastidyl inositol (PI) turnover, hydrolysis, formation of IP_3, calcium mobilization by agonists and phosphorylation of 20 KDa and 47 KDa proteins. Although epinephrine has been shown to facilitate a variety of events associated with platelet activation, its ability to restore the sensitivity of refractory platelets to the action of agonists may not require all the events represented in this scheme.

Oxidation of Epinephrine (Epi) and Concurrent Loss of Its Ability to Induce Platelet Aggregation

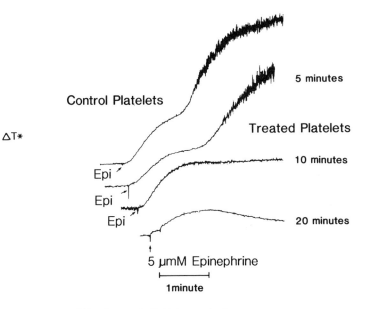

ΔT* change in light transmission

FIGURE 10. Exposure of epinephrine to a constant 6-volt electric source on an electrostat induces time-dependent oxidation. Concurrent with its oxidation, epinephrine loses the ability to cause human platelet aggregation.

desensitize it for further challenge by epinephrine.[103] Radiolabeled epinephrine binding with specific platelet membrane proteins has been reported earlier.[127]

Alteration in the number of alpha receptors on human platelets has been reported under a variety of drug-induced conditions and disease states.[81,120] To determine if the alterations in receptor density occurred during the drug-induced refractory state following epinephrine-induced resensitization, ^{3}H yohimbine binding in three well-characterized refractory models (aspirin, PGE_1, Quin 2-exposed cells) was followed. The number of alpha receptors on aspirin-treated platelets was 280 ± 32, and in those treated with PGE_1 and Quin 2, the numbers were 255 ± 24 and 275 ± 24, respectively. Receptor sites on these drug-induced refractory platelets after exposure to the influence of epinephrine were, aspirin, 237 ± 26; PGE_1, 227 ± 22; and Quin 2, 249 ± 22. Epinephrine-induced restoration of sensitivity to the action of agonists did not alter the number of receptor sites in these refractory platelets. Results of these studies suggest that no significant alteration in receptor density occurs in the drug-induced refractory state or during epinephrine-induced resensitization (Rao and White, unpublished data).

Epinephrine-induced activation has been shown to enhance the formation of binding sites for fibrinogen and to potentiate fibrinogen binding in normal platelets as well as refractory platelets.[4,21,78,112] Studies from our laboratory, using activation-deactivation-reactivation phases of platelets, demonstrated a close correlation between fibrinogen binding and cell activation. Aggregation induced by AA and ADP caused significant binding of radiolabeled fibrinogen. PGI_2-induced disaggregation of platelet clumps resulted in the dissociation of bound fibrinogen. Reactivation and irreversible aggregation due to a challenge by epinephrine and AA or ADP resulted in reassociation of fibrinogen.[21] As discussed in an earlier section, Peerschke[78] also demonstrated that when platelets are stimulated with epinephrine, together with either ADP or thrombin, fibrinogen binding increases by 180% compared to the platelet's response to ADP or thrombin alone. There is considerable published evidence to suggest that epinephrine has a modulating role for fibrinogen binding, and this in turn plays a critical role in platelet aggregation. However, the mechanism by which alpha$_2$-adrenoceptor influences the formation of glycoprotein IIb/IIIa complex or binding sites for the adhesive proteins is not known.

Two well-characterized responses that result from epinephrine-induced platelet activation are promotion of Na^+/H^+ exchange and inhibition of adenylate cyclase. Incubation of refractory platelets with greater than 1 mM amiloride completely blocked the corrective influence of epinephrine. However, this effect of amiloride did not appear specific. Moreover, a recent study has shown that amiloride analogs cause platelet activation and secretion.[132] Because several laboratories are pursuing the role of Na^+/H^+ exchange in platelet activation, studies in our laboratory did not focus on this early event in epinephrine-induced stimulation. Attempts to show that cyclase inhibition may regulate membrane modulation by using adenylate cyclase inhibitors such as 2',5'-dideoxy adenosine and SQ 22536 failed because no potentiating effect by these inhibitors in our drug-induced refractory models could be demonstrated. Similar findings on the use of adenylate cyclase inhibitors were reported earlier.[36]

Since the majority of studies postulate that cytosolic calcium mobilization is critical for platelet activation, studies in our laboratory explored the source and role of calcium in membrane modulation. Using Quin 2-free acid on the outside to buffer ionized calcium, a dose-dependent inhibition in agonist-induced platelet aggregation was demonstrated. Chelation of membrane-associated calcium with CTC also exerted a dose-dependent inhibition of agonist-induced platelet aggregation. Platelet exposure to excess Quin 2-AM and chelation of cytosolic calcium induced inhibition of platelet aggregation. Yet despite the way calcium chelation was achieved in these studies, epinephrine restored the sensitivity of refractory platelets to the action of agonists.[87]

Using Quin 2-loaded platelets, PI metabolism, calcium mobilization, and protein phosphorylation were followed. Quin 2-AM at or above 40 μM concentration (> 4 mM cytosolic concentration) effectively blocked AA acid-induced aggregation. These platelets did not aggregate in response to the action of epinephrine. Yet, a combination of epinephrine and AA caused irreversible aggregation. Using (^{3}H-inositol) radiolabeled platelets, it was demonstrated that a combination of these agents neither enhanced PI hydrolysis nor promoted the formation of increased inositol 1,4,5-triphosphate (IP_3).

Studies with Fura 2-loaded platelets showed that increases in Quin 2 concentration effectively blocked calcium mobilization by AA and epinephrine. Phosphorylation of myosin light chain (20 kDa protein) was considerably inhibited by chelation of cytosolic calcium. Stimulation of Quin 2-loaded platelets induced some degree of phosphorylation of these proteins. However, epinephrine and AA together did not enhance the phosphorylation of this protein to a level equal to or greater than that achieved by those agonists in untreated control platelets. Chelation of cytosolic calcium had minimum effect on PKC-mediated phosphorylation of 47 kDa protein. AA alone could induce phosphorylation of this protein in Quin 2-loaded platelets, although no aggregation occurred. Epinephrine slightly enhanced the phosphorylation of the protein induced by AA.

The specific role of a PKC-dependent pathway in platelet activation is not clear at this writing. Results of these studies suggest that epinephrine-induced restoration of function in Quin 2-loaded AA refractory platelets is not dependent upon accelerated PI metabolism, formation of increased quantities of IP_3, elevation of cytosolic calcium, or enhanced phosphorylation of the myosin light chain.[87]

In spite of concerted efforts by several laboratories to explain how epinephrine induces its facilitating effect on platelet function, the specific mechanism by which it operates is still not clear. What is known, at best, is a series of events that follows receptor-mediated activation.

Further studies must provide a rationale for these events and conclusively demonstrate how each dictates or complements the other. Epinephrine-induced membrane modulation is blocked by yohimbine, a specific alpha$_2$-adrenoceptor blocker, but it is also prevented by partial agonists such as clonidine, phenylephrine, ephedrine and a variety of drugs such as ethanol, verapamil and quinine.[13,86,96] Alpha$_2$-adrenoceptor modulation leads to Na^+/H^+ exchange and alkalinization. The exact role these events play in membrane modulation is not clear. In other cells, increased pH has been postulated to facilitate the opening of calcium channels. Not much is known about voltage-gated or other types of calcium channels in platelets. Increased cytosolic pH may alter substrate specificity for the action of PKC. This enzyme is multifunctional and capable of modulating the activity of other kinases. The 47 kDa protein has been used extensively to monitor the activity of PKC, but not much is known about other substrates that may undergo phosphorylation during platelet activation. Some of the studies discussed earlier[113,114] showed a role for PKC activation in calcium mobilization. However, our studies, using the same inhibitor for PKC (trifluoperazine, 10 μM), demonstrated the opposite effect. Inhibition of PKC resulted in impaired phosphorylation of 47 kDa protein. Treated platelets mobilized significant quantities of calcium and released granule contents when exposed to subthreshold concentrations of thrombin.[84] Platelet aggregation was minimal under these experimental conditions.

We postulate, based on our results, that inhibition of PKC may impair fibrinogen binding and promote accumulation of IP_3. Therefore, the PKC pathway may play a prominent role in the activation of platelets leading to the development of stickiness, whereas the PLC-dependent pathway leading to IP_3 formation may play a critical role in calcium mobilization, contraction-secretion coupling.

In view of the heavy emphasis placed on the central role for calcium as the final common pathway for cell activation, most studies have focused their efforts to demonstrate epinephrine's ability to potentiate events leading to calcium mobilization by various agonists. The internal pool of calcium, although critical to dynamic processes such as contraction-secretion coupling and clot retraction, is not essential for shape change, development of stickiness, activation by artificial surfaces, spreading or irreversible aggregation.[103] Therefore, in drug-induced refractory platelets, epinephrine-induced membrane modulation may not modulate or promote any events related to PI metabolism or calcium mobilization. It may operate strictly at the surface membrane level and promote IIb/IIIa complex formation, induce phosphorylation of proteins via stimulation of kinases and, thereby, play a role in the development of stickiness.

Very little is known at this writing about the specific mechanism involved in the development of stickiness. There are many unanswered questions, but many pieces of the puzzle have been discovered. In spite of the contradictions and conflicting reports, it is reasonable to conclude that epinephrine, on its own merit, induces activation of $alpha_2$-adrenoceptors, promotes Na^+/H^+ exchange, enhances alkalinization of cytosolic pH, and potentiates the action of physiological agonists in modulating biochemical events following activation.

CLINICAL SIGNIFICANCE OF MEMBRANE MODULATION

Catecholamines have been shown to play a modulatory role in the alteration of bleeding times.[5,24,42,63,68,75] O'Brien,[71] using phentolamine as an antagonist, demonstrated the influence of epinephrine in lowering bleeding time. Stormorken and associates[118] described a bleeding disorder in which platelets of the patient failed to aggregate in response to epinephrine, although no receptor deficiency could be demonstrated. Similarly, in myeloproliferative disease, platelets have been known to have a normal component of $alpha_2$-adrenoceptors and yet be functionally insensitive to epinephrine.[122] Plasma catecholamines are known to modulate $alpha_2$-adrenoceptor sensitivity in hypertensive individuals[40] and in congestive heart failure,[130] whereas increased adrenal tone has been implicated in the pathogenesis of high blood pressure, platelet activation in pre-eclampsia,[52] acute myocardial infarction and angina pectoris.[1,35,110]

Enhanced affinity or sensitivity to epinephrine has been suggested as a possible mechanism of platelet hypersensitivity in acute myocardial ischemia[65] and in variant angina.[131] Recent epidemiological studies have reported a circadian variation in the frequency of acute myocardial infarction onset[70] and sudden cardiac death.[69] Diurnal variations in platelet stickiness, concurrent morning increase in epinephrine levels, and platelet aggregability have been suggested as risk factors for myocardial infarction and sudden death.[5,126]

The platelet release reaction was considered essential for irreversible aggregation and formation of thrombi.[41] Discovery of the novel metabolites of AA, including PGI_2 and thromboxane, suggested a modulating role for these bioactive lipids in platelet endothelial interaction.[62] Based on this knowledge, drug trials were designed to reduce clinical implications associated with ischemic heart disease using cyclooxygenase inhibitors, particularly aspirin. However, ingestion of this drug will not totally inhibit platelet function, and the inhibitory effect can be reversed by epinephrine. Therefore, it is essential to find specific agents that can antagonize epinephrine's effect on platelets while still maintaining normal hemostasis.

In a preliminary study, antagonists such as yohimbine, atenolol, verapamil, and ethanol were tested for their ability to potentiate the action of low aspirin on platelet function. Aspirin ingestion alone did not prevent epinephrine-induced restoration of sensitivity to the action of AA. However, both yohimbine and atenolol were effective in blocking epinephrine-induced restoration of sensitivity to the action of agonists.[86]

In an earlier study, a similar effect of clonidine and yohimbine on the influence of epinephrine at a single cell level was demonstrated.[115] The novel mechanism of membrane modulation, on the other hand, may be the only salvage mechanism available for restoration of function in drug-induced refractory platelets and in platelets of patients with various bleeding disorders. It is reasonable, therefore, to speculate that this intrinsic mechanism may be responsible for normal hemostasis in situations where platelets in circulation have compromised function. Similar to the reported action of epinephrine, recent studies have shown desmopressin (DDAVP) to be effective in reducing prolonged bleeding time in patients with congenital platelet function defects.[108]

SUMMARY

Platelets play an important role in the maintenance of normal hemostasis and in the initiation of thrombosis. Biochemical events associated with the lifesaving function of platelets are the same as those involved in the initiation of pathological processes. Epinephrine has been shown to potentiate all of the biochemical events associated with platelet activation and to modulate functions such as shape change, adhesion, secretion of granule contents and irreversible aggregation. This hormone induces its effect on platelets by acting through the $alpha_2$-adrenoceptor. Activation of this receptor has been shown to induce

Na$^+$/H$^+$ exchange and increase cytosolic pH. Stimulation of multifunctional PKC seems to facilitate the formation of binding sites for adhesive proteins by an as yet unknown mechanism. Glycoprotein IIb/IIIa complex formation has been shown to promote fibrinogen binding. Elevation of cytosolic calcium seems to facilitate secretion of granule contents.

A series of studies demonstrated the existence of an intrinsic mechanism termed "membrane modulation," capable of securing irreversible aggregation of platelets with compromised function. This newly discovered mechanism is alpha$_2$-adrenoceptor dependent, and does not require the mediation of prostanoids, ADP or PAF. It is dependent upon the availability of calcium, and enhances calcium uptake by the plasma membranes, potentiates fibrinogen binding, and promotes phosphorylation of proteins. Epinephrine-induced membrane modulation is not dependent upon enhanced phosphoinositol metabolism, elevation of cytosolic calcium, or phosphorylation of myosin light chain. It is a novel salvage mechanism capable of restoring platelet sensitivity to agonists when the platelet's function is compromised.

ACKNOWLEDGMENTS

The author thanks Ms. Susan Schwarze and Ms. Kay Dressler for their help in preparing this manuscript. This work was supported by USPHS grants HL-11880, CA-21767. GM-22167 and HL-30217, a grant from the March of Dimes (MOD1-886), CRC program RR400, the Minnesota Medical Foundation (HDRF 38-45, HRF 47-87, HRF 57-88) and the Graduate School. This article is not a review on epinephrine and platelet function; therefore, we apologize to many researchers in this field if their valuable contribution is not cited.

List of Abbreviations

5-HT = serotonin
AA = arachidonate
ADP = adenosine diphosphate
cAMP = cyclic adenosine monophosphate
CTC = chlortetracycline
DDAVP = desmopressin
ETYA = 5,8,11,14 eicosatetraynoic acid
HPS = Hermansky-Pudlak syndrome
IP_3 = inositol 1,4,5-triphosphate
kDa = kilodalton
LA = phospholipase A_2
PAF = platelet activating factor
PGD_2 = prostaglandin D_2
PGE_1 = prostaglandin E_1
PGI_2 = prostacyclin
PI = phosphastidyl inositol
PIP = phosphatidylinositol 4-phosphate
PIP_2 = phosphatidyl 4,5-biphosphate
PKC = protein kinase C
PLC = phospholipase C
SPD = storage pool disorder
U46619 = endoperoxide mimetic
U53119 = 4,7,10,13-eicosatetraynoic acid

REFERENCES

1. **Ardlie, N.G., Cameron, H.A., and Garrett, J.,** Platelet activation by circulating levels of hormones: A possible link in coronary heart disease, *Thromb. Res.,* 36, 315-322, 1984.
2. **Banga, H.S., Simons, E.R., Brass, L.F., and Rittenhouse, S.E.,** Activation of phospholipase A and C in human platelets exposed to epinephrine: Role of glycoproteins IIb/IIIa and dual role of epinephrine, *Proc. Natl. Acad. Sci. USA,* 83, 9197-9201, 1986.
3. **Beatty, C.H., Howard, C.F., and Caruso, V.,** Potentiation with epinephrine of Macaque platelet aggregation by other agonists: Implications for studies on human atherosclerosis, *Thromb. Res.,* 41, 447-458, 1986.
4. **Bennett, J.S., and Vilaire, G.,** Exposure of platelet fibrinogen receptors by ADP and epinephrine, *J. Clin. Invest.,* 64, 1393-1401, 1979.
5. **Besterman, E., Myat, G., and Trivadi, V.,** Diurnal variation of platelet stickiness compared with effects of adrenaline, *Br. Med. J.,* 1, 597-600, 1967.
6. **Bolli, R., Ware, J.A., Brandon, T.A., Weilbaecher, D.G., and Mace, M.L.,** Platelet-mediated thrombosis in stenosed canine coronary arteries: Inhibition by nicergoline, a platelet-active, alpha-adrenergic antagonist, *J. Am. Cardiol.,* 3, 1417-1426, 1984.
7. **Born, G.V.R., and Cross, M.J.,** The aggregation of blood platelets, *J. Physiol.,* 168, 178-195, 1963.
8. **Braquet, P., Touqui, L., Shen, T.Y., and Vergaftig, B.B.,** Perspectives in platelet-activating factor research, *Pharmacol. Rev.,* 39, 97-145, 1987.
9. **Brass, L.F., and Shattil, S.J.,** Changes in surface-bound and exchangeable calcium during platelet activation, *J. Biol. Chem.,* 257, 14000-14005, 1982.

10. **Burris, S.M., Smith, C.M., Rao, G.H.R., and White, J.G.,** Aspirin treatment reduces platelet resistance to deformation, *Arteriosclerosis,* 7, 385-388, 1987.

11. **Bushfield, M., McNicol, A., and MacIntyre, D.E.,** Possible mechanisms of the potentiation of blood platelet activation by adrenaline, *Biochem. J.,* 241, 671-676, 1987.

12. **Cameron, H.A., and Ardlie, N.G.,** The facilitating effects of adrenaline on platelet aggregation, *Prostaglandins Leukotrienes Med.,* 9, 117-128, 1982.

13. **Charney, D.S., Heninger, H.G.S., and Sternberg, D.E.,** Assessment of α_2 adrenergic autoreceptor function in humans: Effects of oral yohimbine, *Life Sci.,* 30, 2033-2041, 1982.

14. **Chignard, M., LeCouedic, J.P., Tence, M., Vargaftig, B.B., and Benveniste, J.,** The role of platelet activating factor in platelet aggregation, *Nature,* 275, 799-800, 1979.

15. **Chignard, M., LeCouedic, J.P., Vargaftig, B.B., and Benveniste, J.,** Platelet activating factor (PAF-acether) secretion from platelets: Effects of aggregating agents, *Br. J. Haematol.,* 46, 455-464, 1980.

16. **Clare, K.A., and Scrutton, M.C.,** The role of Ca^{2+} uptake in the response of human platelets to adrenaline and to 1-0-alkyl-2-acetyl-8n-glycero-3-phosphocholine (platelet-activating-factor), *Eur. J. Biochem.,* 140, 129-136, 1984.

17. **Connolly, T.M., and Limbird, L.E.,** Removal of extra platelet Na^+ eliminates indomethacin-sensitive secretion from human platelets stimulated by epinephrine and thrombin, *Proc. Natl. Acad. Sci. USA,* 80, 5320-5324, 1983.

18. **Connolly, T.M., and Limbird, L.E.,** The influence of Na^+ on the alpha$_2$-adrenergic receptor system of human platelets. A method for removal of extra platelet Na^+. Effect of Na^+ removal on aggregation, secretion and cAMP accumulation, *J. Biol. Chem.,* 258, 3907-3912, 1983.

19. **Corby, D.G., and O'Barr, T.P.,** Decreased alpha-adrenergic receptors in newborn patients: Cause of abnormal response to epinephrine, *Dev. Pharmacol. Ther.,* 2, 215-225, 1981.

20. **Cowen, P., Elliott, J.M., Fraser, S., and Stump, K.,** Effect of blood loss on platelet A$_2$-adreno receptor, *Br. J. Pharmacol., Supp.* (1), 80, 460, 1983.

21. **Cox, A.C., Carroll, R.C., White, J.G., and Rao, G.H.R.,** Recycling of platelet phosphorylation and cytoskeletal assembly, *J. Cell Biol.,* 98, 8-15, 1984.

22. **Crouch, M.F., and Lapetina, E.G.,** A role for G_i in control of thrombin receptor-phospholipase C coupling in human platelets, *J. Biol. Chem.,* 263, 3363-3371, 1988.

23. **de Chaffoy de Courcelles, D., Rovens, P., Van Belle, H., and de Clerk, F.,** The synergistic effect of serotonin and epinephrine on the human platelet at the level of signal transduction, *FEBS Lett.,* 219, 283-288, 1987.

24. **DeRouaux, G.,** Action des amines sympathicometianes sur l'hemostase spontanee, *Arch. Int. Pharmacodyn.,* LXV, 125-195, 1941.

25. **Didisheim, P., and Fuster, V.,** Actions and clinical status of platelet suppressive agents, *Semin. Hematol.,* 15, 55-72, 1978.

26. **Doyle, M.C., George, A.J., Ravindran, A.V., and Philpott, R.,** Platelet alpha$_2$-adrenoreceptor binding in elderly depressed patients, *Am. J. Psychiatry,* 142, 1489-1490, 1985.

27. **Figures, W.R., Searce, L.M., Wachtfogel, Y., Chen, J., Colman, R.F., and Colman, R.W.,** Platelet ADP receptor and alpha$_2$-adrenoreceptor interaction. Evidence for an ADP requirement for epinephrine-induced platelet activation and an influence of epinephrine on ADP-binding, *J. Biol. Chem.,* 261, 5981-5986, 1986.

28. **Folts, J.D., and Gallagher, K.P.,** Epinephrine-induced platelet aggregation in mechanically stenosed dog coronary arteries occurring after aspirin administration, *Clin. Res.,* 26, 231A, 1978.

29. **Folts, J.D., Rowe, G.G., and Rao, G.H.R.,** Problems with aspirin as antithrombotic agent in coronary heart disease, *Lancet,* 1, 937-938, 1988.

30. **Folts, J.D., and Smith, S.O.,** Platelet thrombus formation in stenosed monkey carotid arteries: Inhibited with aspirin, restored with epinephrine, *J. Am. Cardiol.,* 5, 468A, 1985.

31. **Fouque, F., and Vargaftig, B.B.,** Triggering by PAF-acether and adrenaline of cyclo-oxygenase-independent platelet aggregation, *Br. J. Pharmacol.,* 83, 625-633, 1984.

32. **Fuster, V., and Chesebro, J.H.,** Antithrombotic therapy: Role of platelet-inhibitor drugs II. Pharmacologic effects of platelet-inhibitor drugs, *Mayo Clin. Proc.,* 56, 185-195, 1981.

33. **Gaarder, A., Jonsen, J., Lalland, S., Hellem, A., and Owen, P.A.,** Adenosine diphosphate in red cells as a factor in adhesiveness of human blood platelets, *Nature,* 192, 531-532, 1961.

34. **Garcia-Sevilla, J.A., Ugedo, L., Ulibarri, I., and Gutierrez, M.,** Heroin decreases the density of platelet A_2-adrenoreceptors in human addicts, *Psychopharmacology,* 88, 489-492, 1986.

35. **Gazes, P.C., Richardson, J.A., and Woods, E.F.,** Plasma catecholamine concentrations in myocardial infarction and angina pectoris, *Circulation,* 19, 657-661, 1959.

36. **Grant, J.A., and Scrutton, M.C.,** Interaction of selective A-adrenoreceptor agonists and antagonists with human and rabbit blood platelets, *Br. J. Pharmacol.,* 71, 121-134, 1980.

37. **Hallam, T.J., Scrutton, M.C., and Wallis, R.B.,** Responses of rabbit platelets to adrenaline induced by other agonists, *Thromb. Res.,* 20, 413-424, 1981.

38. **Haslam, R.J.,** Role of adenosine diphosphate in the aggregation of blood-platelets by thrombin and by fatty acids, *Nature,* 202, 765-768, 1964.

39. **Hoffman, B.B., and Lefkowitz, R.J.,** Alpha-adrenergic receptor subtypes, *N. Engl. J. Med.,* 302, 1390-1396, 1980.

40. **Hollister, A.S., Onrot, J., Lonce, S., Nadeau, J.H.J., and Robertson, D.,** Plasma catecholamine modulation of alpha$_2$-adrenoreceptor agonist affinity and sensitivity and hypertensive human platelets, *J. Clin. Invest.,* 77, 1416-1421, 1986.

41. **Holmsen, H.,** Prostaglandin-thromboxane synthesis and secretion as positive feedback loops in the propagation of platelet responses, *Thromb. Haemost.,* 38, 1030-1042, 1977.

42. **Honour, A.J., and Mitchell, J.R.A.,** Platelet clumping *in vivo, Nature,* 197, 1019-1020, 1963.

43. **Huang, E.M., and Detwiler, T.C.,** Characteristics of the synergistic actions of platelet agonists, *Blood,* 57, 685-690, 1981.

44. **Johnson, G.J., Leis, L.A., Rao, G.H.R., and White, J.G.,** Arachidonate-induced platelet aggregation in the dog, *Thromb. Res.,* 14, 147-154, 1979.

45. **Johnson, G.J., Rao, G.H.R., Leis, L.A., and White, J.G.,** Effects of agents which alter cyclic AMP on arachidonate-induced platelet aggregation in the dog, *Blood,* 55, 722-729, 1980.

46. **Johnson, G.J., Rao, G.H.R., and White, J.G.,** Epinephrine potentiation of thromboxane-stimulated platelet aggregation, *Clin. Res.,* 28, 729A, 1980.

47. **Johnson, G.J., Rao, G.H.R., and White, J.G.,** Epinephrine potentiates thromboxane A_2-stimulated human platelet aggregation by an alpha adrenergic receptor-linked mechanism, *Clin. Res.,* 29, 236A, 1981.

48. **Jorgensen, L., Rowsell, H.C., Hovig, T., Glynn, M.F., and Mustard, J.F.,** Adenosine diphosphate-induced platelet aggregation and myocardial infarction in swine, *Lab. Invest.,* 17, 616-644, 1967.

49. **Kawahara, Y., Yamanishi, J., Tsunemitsu, M., and Fukuzaki, H.,** Protein phosphorylation and diglyceride production during serotonin release induced by epinephrine plus ADP in human platelets, *Thromb. Res.,* 30, 477-485, 1983.

50. **Kerry, R., and Scrutton, M.C.,** Platelet adrenoreceptors, in *The Platelets: Physiology and Pharmacology,* Longenecker, G.L., Ed., Academic Press, Inc., Orlando, Florida, 1985, 113-159.

51. **Kinlough-Rathbone, R., Packham, M.A., and Mustard, J.F.,** Synergism between platelet aggregating agents: The role of the arachidonate pathway, *Thromb. Res.,* 11, 567-580, 1977.

52. **Kjeldsen, S.E., Eide, I. Akesson, I., Oian, P., Maltau, J.M., Lande, K., and Gjesdal, K.,** Increased arterial adrenaline is highly correlated to blood pressure and *in vivo* platelet functioning pre-eclampsia, *J. Hypertens.,* 3, 593-595, 1985.

53. **Kobilka, B.K., Kobilka, T.S., Daniel, K., Regan, J.W., Caron, M.G., and Lefkowitz, R.J.,** Chimeric A_2-B$_2$-adrenergic receptors: Delineation of domains involved in effector coupling and ligand binding specificity, *Science,* 240, 1310-1316, 1988.

54. **Kobilka, B.K., Matsui, H., Kobilka, T.S., Yang-Feng, T.L., Francke, U., Caron, M.G., Lefkowitz, R.J., and Regan, J.W.,** Cloning, sequencing, and expression of the gene loading for the human platelet A_2-adrenoreceptor, *Science,* 238, 650-656, 1987.

55. **Lalau-Keraly, C., Vickers, J.D., Kinlough-Rathbone, R.L., and Mustard, J.F.,** Involvement of phosphoinositide metabolism in potentiation by adrenaline of ADP-induced aggregation of rabbit platelets, *Biochem. J.,* 242, 841-847, 1987.

56. **Lanza, F., Cazenave, J., Beretz, A., Sutter-Bay, A., Kretz, J., and Kieny, R.,** Potentiation by adrenaline of human platelet activation and the inhibition by the alpha-adrenergic antagonist nicergoline of platelet adhesion. Secretion and aggregation, *Agents Actions,* 18, 586-595, 1986.

57. **Lefkowitz, R.J., and Caron, M.C.,** Adrenergic receptors: Models for the study of receptors coupled to guanine nucleotides regulatory proteins, *J. Biol. Chem.,* 263, 4993-4996, 1988.

58. **Lehmann, M., Hasler, K., Bergdolt, E., and Keul, J.,** Alpha-2-adrenoreceptor density on intact platelets and adrenaline-induced platelet aggregation in endurance and non-endurance-trained subjects, *Int. J. Sports Med.,* 7, 172-176, 1986.

59. **Levine, S.P., Towell, B.P., Suarez, A.M., Knieriem, L.K., Harris, M.M., and George, J.N.,** Platelet activation and secretion associated with emotional stress, *Circulation,* 71, 1129-1134, 1985.

60. **Limbird, L.E.,** Receptors linked to inhibition of adenylate cyclase: Additional signaling mechanisms, *FASEB J.,* 2, 2686-2695, 1988.

61. **Limbird, L.E., Connolly, T.M., Sweatt, J.D., and Uderman, H.D.,** Human platelet alpha$_2$ adrenergic receptors: Effect of Na^+ on interaction with the adenylate cyclase system and on epinephrine-stimulated platelet secretion, *Adv. Cyclic Nucleotide Protein Phosphorylation Res.,* 19, 235-242, 1985.

62. **Marcus, A.J.,** The role of lipids in platelet function: With particular reference to the arachidonic acid pathway, *J. Lipid Res.,* 19, 793-826, 1978.

63. **McClure, P.D., Ingram, G.I.C., and Jones, R.V.,** Platelet changes after adrenaline blockers. *Thromb. Diath. Haemorrh.,* 13, 136-139, 1965.

64. **McGrath, J., and Wilson, V.,** A-adrenoreceptor subclassification by classical and response-related methods: Same question, different answers, *Trends Pharmacol. Sci.,* 9, 162-165, 1988.

65. **Mehta, J., Mehta, P., and Ostrowski, N.,** Increase in human platelet A_2-adrenergic receptor affinity for agonist in unstable angina, *J. Lab. Clin. Med.,* 106, 661-666, 1985.

66. **Mills, D.C.B., and Roberts, G.C.K.,** Effects of adrenaline on human blood platelets, *J. Physiol.,* 193, 443-453, 1967.

67. **Milton, J.G., and Frojmovic, M.M.,** Adrenaline and adenosine diphosphate-induced platelet aggregation require shape change. Importance of pseudopods, *J. Lab. Clin. Med.,* 104, 805-815, 1984.

68. **Mitchell, J.R.A., and Sharp, A.A.,** Platelet clumping *in vitro, Br. J. Haematol.,* 10, 78-93, 1964.

69. **Müller, J.E., Ludmer, P.L., Willich, S.N., Tofler, G.H., Aylmer, G., Klangos, I., and Stone, P.H.,** Circadian variation in the frequency of sudden cardiac death, *Circulation,* 75, 131-138, 1987.

70. **Müller, J.E., Stone, P.H., Turi, Z.G., Rutherford, J.D., Czeisler, C.A., Parker, C., Poole, W.K., Passamani, E., Roberts, R., Robertson, T., Sobel, B.E., Willerson, J.T., Braunwauld, E. and the Millis Study Group,** Circadian variation in the frequency of onset of acute myocardial infarction, *N. Engl. J. Med.,* 313, 1315-1322, 1985.

71. **O'Brien, J.R.,** Some effects of adrenaline and anti-adrenaline compounds in platelets *in vitro* and *in vivo, Nature,* 200, 763-764, 1963.

72. **Otto-Erich, B., Anlauf, M., Graven, N., and Bock, K.D.,** Age-dependent decrease of A_2-adrenergic receptor number in human platelets, *Eur. J. Pharmacol.,* 81, 345-347, 1982.

73. **Owen, N.E., Feinberg, H., and LeBreton, G.C.,** Epinephrine induces Ca^{2+} uptake in human blood platelets, *Am. J. Physiol.,* 239, H483-H488, 1980.

74. **Owen, N.E., and LeBreton, G.C.,** The involvement of calcium in epinephrine or ADP potentiation of human platelet aggregation, *Thromb. Res.,* 17, 855-863, 1980.

75. **Ozge, A.H., Mustard, J.F., Hegardt, B., Roswell, H.C., and Downie, H.G.,** The effect of adrenaline on blood coagulation, platelet economy and thrombus formation, *Can. Med. Assoc. J.,* 88, 265, 1963.

76. **Packham, M.A., and Mustard, J.F.,** Pharmacology of platelet-affecting drugs, *Circulation,* 62, 26-41, 1980.

77. **Patscheke, H.,** Role of activation in epinephrine-induced aggregation of platelets, *Thromb. Res.,* 17, 133-142, 1980.

78. **Peerschke, E.I.B.,** Effect of epinephrine on fibrinogen receptor exposure by aspirin-treated platelets and platelets from concentrates in response to ADP and thrombin, *Am. J. Hematol.,* 16, 335-345, 1984.

79. **Peterson, D.A., and Gerrard, J.M.,** Reduction of a disulfide bond by beta-adrenergic agonists: Evidence in support of a general "reductive activation" hypothesis for the mechanism of action of adrenergic agents, *Med. Hypotheses,* 22, 45-49, 1987.

80. **Peterson, D.A., Gerrard, J.M., Glover, S.M., Rao, G.H.R., and White, J.G.,** Epinephrine reduction of Heme: Implication for understanding the transmission of an agonist stimulus, *Science,* 215, 71-73, 1982.

81. **Piletz, J.E., Schubert, D.S.P., and Halaris, A.,** Evaluation of studies on platelet alpha$_2$ adrenoreceptors in depressive illness, *Life Sci.,* 39, 1589-1616, 1986.

82. **Pogliani, E., A. Della Volpe, R. Ferrari, P. Recalcati, and C. Praga,** Inhibition of human platelet aggregation by oral administration of nicergoline: A double blind study, *Farmaco,* 30, 630-635, 1975.

83. **Powling, M.J., and Hardisty, R.M.,** Potentiation by adrenaline of Ca^{2+} influx and mobilization in stimulated human platelets: Dissociation from thromboxane generation and aggregation, *Thromb. Haemost.,* 59, 212-215, 1988.

84. **Rao, G.H.R.,** Influence of calmodulin antagonist (Stelazine) on agonist-induced calcium mobilization and platelet activation, *Biochem. Biophys. Res. Commun.,* 148, 768-775, 1987.

85. **Rao, G.H.R., Escolar, G., and White, J.G.,** Epinephrine reverses the inhibitory influence of aspirin on platelet vessel wall interaction, *Thromb. Res.,* 44, 65-74, 1986.

86. **Rao, G.H.R., Escolar, G., Zavrol, J., and White, J.G.,** Influence of adrenergic receptor blockade on aspirin-induced inhibition of platelet function, *Platelets,* 1, 145-150, 1990.

87. **Rao, G.H.R., Gerrard, J.M., Cohen, I., Witkop, C.J., and White, J.G.,** Origin and role of calcium in platelet activation—A contraction-secreting coupling, in *Cell Calcium Metabolism,* Fiskum, G., Ed., Plenum Press, New York, 1989, 411-427.

88. **Rao, G.H.R., Gerrard, J.M., and White, J.G.,** Epinephrine-induced potentiation of arachidonate aggregation in Quin 2-loaded platelets is not mediated by elevation of cytosolic calcium, *Thromb. Haemost.,* 58, 962A, 1987.

89. **Rao, G.H.R., Gerrard, J.M., Witkop, C.J., and White, J.G.,** Platelet aggregation independent of ADP release on prostaglandin synthesis in patients with the Hermansky-Pudlak syndrome, *Prostaglandins Med.,* 6, 459-472, 1981.

90. **Rao, G.H.R., Johnson, G.J., and White, J.G.,** Influence of epinephrine on the aggregation response of aspirin-treated platelets, *Prostaglandins Med.,* 5, 45-58, 1980.

91. **Rao, G.H.R., Peller, J.D., Semba, C.P., and White, J.G.,** Influence of the calcium sensitive fluorophore Quin 2 on platelet function, *Blood,* 67, 354-61, 1986.

92. **Rao, G.H.R., Radha, E., and White, J.G.,** Irreversible platelet aggregation does not depend on lipoxygenase metabolites, *Biochem. Biophys. Res. Commun.,* 131, 50-57, 1985.

93. **Rao, G.H.R., Reddy, K.R., and White, J.G.,** Influence of trifluoperazine on platelet aggregation and disaggregation, *Prostaglandins Med.,* 5, 221-234, 1980.

94. **Rao, G.H.R., Reddy, K.R., and White, J.G.,** The influence of epinephrine on prostacyclin (PGI$_2$) induced dissociation of ADP aggregated platelets, *Prostaglandins Med.,* 4, 385-397, 1980.

95. **Rao, G.H.R., Reddy, K.R., and White, J.G.,** Modification of human platelet response to sodium arachidonate by membrane modulation, *Prostaglandins Med.,* 6, 75-90, 1981.

96. **Rao, G.H.R., Reddy, K.R., and White, J.G.,** Low dose aspirin, platelet, function and prostaglandin synthesis: Influence of epinephrine and alpha adrenergic blockade, *Prostaglandins Med.,* 6, 485-494, 1981.

97. **Rao, G.H.R., Reddy, K.R., and White, J.G.,** Penicillin induced human platelet dysfunction and its reversal by epinephrine, *Prostaglandins Leukotrienes Med.,* 11, 199-211, 1983.

98. **Rao, G.H.R., Schmid, H.H.O., Reddy, K.R., and White, J.G.,** Human platelet activation by an alkyl-acetyl analogue of phosphatidylcholine, *Biochim. Biophys. Acta,* 715, 205-214, 1982.

99. **Rao, G.H.R., and White, J.G.,** Epinephrine potentiation of arachidonate-induced aggregation of cyclooxygenase deficient platelets, *Am. J. Hematol.,* 11, 355-366, 1981.

100. **Rao, G.H.R., and White, J.G.,** Platelet activating factor (PAF) causes human platelet aggregation through the mechanism of membrane modulation, *Prostaglandins Leukotrienes Med.,* 9, 459-472, 1982.

101. **Rao, G.H.R., and White, J.G.,** Disaggregation and reaggregation of "irreversibly" aggregated platelets. A method for more complete evaluation of antiplatelet drugs, *Agents Actions,* 16, 425-433, 1985.

102. **Rao, G.H.R., and White, J.G.,** Role of arachidonic acid metabolism in human platelet activation and irreversible aggregation, *Am. J. Hematol.,* 19, 339-347, 1985.

103. **Rao, G.H.R., and White, J.G.,** Influence of phospholipase A_2 on human blood platelet alpha adrenergic receptor function, *Thromb. Res.,* 53, 427-434, 1989.

104. **Rao, G.H.R., and White, J.G.,** Aspirin, PGE, and Quin-2 AM induced platelet dysfunction. Restoration of function by norepinephrine, *Prost. Leukot. Essen. Fatty Acids,* 39, 141-146, 1990.

105. **Rossi, E.C., Louis, G., and Zeller, E.A.,** Structure activity relationships between catecholamines and the A-adrenergic receptor responsible for the aggregation of human platelets by epinephrine, *J. Lab. Clin. Med.,* 93, 286-294, 1979.

106. **Sakaguchi, K., Hattori, R., Yui, Y., Takatsu, Y., Susawa, T., Yui, N., Nonogi, H., Tamaki, S., and Kawai, C.,** Altered platelet alpha$_2$ adrenoreceptor in acute myocardial infarction and its relation to plasma catecholamine concentrations, *Br. Heart J.,* 55, 434-438, 1986.

107. **Salzman, E.W., Chambers, A.D., and Neri, L.L.,** Possible mechanism of aggregation of blood platelets by adenosine diphosphate, *Nature,* 210, 167-169, 1966.

108. **Schulman, S., Johnsson, H., Egelsberg, N., and Blombach, M.,** DDAVP-induced correction of prolonged bleeding time in patients with congenital platelet function defects, *Thromb. Res.,* 45, 165-174, 1987.

109. **Seiss, W., Weber, P.C., and Lapetina, E.G.,** Activation of phospholipase C is dissociated from arachidonate metabolism during platelet shape change induced by thrombin or platelet activating factor. Epinephrine does not induce phospholipase C activation on platelet shape changes, *J. Biol. Chem.,* 259, 8286-8292, 1984.

110. **Seitz, R., Leising, H., Lieberman, A., Rohner, I., Cerdes, H., and Egbring, R.,** Possible interaction of platelets and adrenaline in the early pace of myocardial infarction, *Res. Exp. Med. (Berl.),* 187, 385-393, 1987.

111. **Shattil, S.J., and Brass, L.F.,** Induction of fibrinogen receptor on human platelets by intracellular mediators, *J. Biol. Chem.,* 262, 992-1000, 1987.

112. **Shattil, S.J., Motulsky, H.J., Insel, P.A., Flaherty, L., and Brass, L.F.,** Expression of fibrinogen receptors during activation and subsequent desensitization of human platelets by epinephrine, *Blood,* 68, 1224-1231, 1986.

113. **Siffert, W., and Akkerman, J.N.,** Protein kinase C enhances Ca^{2+} mobilization in human platelets by activating Na^+/H^+ exchange, *J. Biol. Chem.,* 263, 4223-4227, 1988.

114. **Siffert, W., and Scheid, P.,** A phorbol ester and 1-oleoyl-2-acetylglycerol induce Na^+/H^+ exchange in human platelets, *Biochem. Biophys. Res. Commun.,* 141, 13-19, 1986.

115. **Smith, C.M., Burris, S.M., Rao, G.H.R., and White, J.G.,** Epinephrine-induced reversal of aspirin effects on platelet deformability, *Thromb. Res.,* 51, 35-44, 1988.

116. **Smith, J.B.,** The prostanoids in hemostasis and thrombosis: A review, *Am. J. Pathol.,* 99, 742-804, 1980.

117. **Steen, V.M., and Holmsen, A.,** Synergism between thrombin and epinephrine in human platelets: Different dose-response relationships for aggregation and dense granule secretion, *Thromb. Haemost.,* 54, 680-683, 1985.

118. **Stormorken, H., Gogstad, G. and Solum, N.O.,** A new bleeding disorder: Lack of platelet aggregatory response to adrenaline and lack of secondary aggregation to ADP and platelet activating factor, *Thromb. Res.,* 29, 391-402, 1982.

119. **Stormorken, H., Lyberg, T., Hakvaag, L., and Nakstad, B.,** Dissociation of the aggregating effect and the inhibitory effect upon cyclic adenosine monophosphate accumulation by adrenaline and adenosine diphosphate in human platelets, *Thromb. Res.,* 45, 363-370, 1987.

120. **Sundareshan, P.R., Weitraub, M., Hershey, L.A., Kroening, B.H., Harday, J., and Banerjee, S.P.,** Platelet alpha-adrenergic receptors in obesity. Alterations with weight loss, *Clin. Pharmacol. Ther.,* 33, 776-785, 1983.

121. **Supiano, M.A., Linares, O.A., Halter, J.B., Reno, K.M., and Rosen, S.G.,** Functional uncoupling of the platelet A_2-adrenergic receptor-adenylate cyclase complex in the elderly, *J. Clin. Endocrinol. Metab.,* 64, 1160-1164, 1987.

122. **Swart, S.S., Pearson, D., Wood, J.K., and Barnett, D.B.,** Human platelet alpha$_2$-adrenoreceptors: Relationship between radioligand studies and adrenaline-induced aggregation in normal individuals, *Eur. J. Pharmacol.,* 103, 25-32, 1984.

123. **Sweatt, J.W., Connolly, T.M., Cragoe, E.J., and Limbird, L.E.,** Evidence that Na^+/H^+ exchange regulates receptor-mediated phospholipase A_2 activation in human platelets, *J. Biol. Chem.,* 261, 8667-8673, 1986.

124. **Thomas, D.P.,** The role of platelet catecholamines in the aggregation of platelets by collagen and thrombin, *Exp. Biol. Med.,* 3, 129-134, 1968.

125. **Thompson, N.T., Scrutton, M.C., and Wallus, R.B.,** Synergistic responses in human platelets. Comparison between aggregation, secretion and cytosolic Ca^{2+} concentration, *Eur. J. Biochem.,* 161, 399-408, 1986.

126. **Tofler, G.H., Brezinski, D., Schafer, A.I., Czeisler, C.A., Rutherford, J.D., Willich, S.N., Gleason, R.E., Williams, G.H., and Müller, J.E.,** Concurrent morning increase in platelet aggregability and the risk of myocardial infarction and sudden cardiac death, *N. Engl. J. Med.,* 316, 1514-1518, 1987.

127. **Vainer, H.,** 14C-epinephrine and interaction with platelet membrane binding sites, *Cell Struct. Funct.,* 2, 267-280, 1977.

128. **Vargaftig, B.B., Fouaue, F., Benveniste, J., and Odiot, J.,** Adrenaline and PAF-acether synergize to trigger cyclooxygenase-independent activation of plasma-free human platelets, *Thromb. Res.,* 28, 557-573, 1982.

129. **Weiss, H.J.,** Antiplatelet therapy: Clinical uses of antiplatelet drugs, *N. Engl. J. Med.,* 298, 1344-1347 and 1413-1416, 1978.

130. **Weiss, R.J., Tobes, M., Wertz, C.E., and Smith, C.B.,** Platelet Alpha$_2$-adrenoreceptors in chronic congestive heart failure, *Am. J. Cardiol.,* 52, 101-105, 1983.

131. **Yokoyama, M., Kawashima, S., Sakamoto, S., Akita, H., Okada, T., Mizutani, T., and Fukuzaki, H.,** Platelet reactivity and its dependence on alpha-adrenergic receptor function in patients with ischemic heart disease, *Br. Heart J.,* 49, 20-25, 1983.

132. **Yoshida, K., and Cragoe, E.J. Jr.,** Amiloride analogs stimulate protein phosphorylation and secretion in human platelets, *Biochem. Int.,* 16, 913-921, 1988.

133. **Zavoico, G.B., and Cragoe, E.J. Jr.** Calcium mobilization can occur independent of acceleration of Na^+/H^+ exchange in thrombin stimulated human platelets, *J. Biol. Chem.,* 263, 9635-9640, 1988.

134. **Zavoico, G.B., Cragoe, E.J., and Feinstein, M.B.,** Regulation of intracellular pH in human platelets. Effects of thrombins A23187, and ionomycin and evidence for activation of Na^+/H^+ exchange and its inhibition by amiloride analogs, *J. Biol. Chem.,* 261, 13160-13167, 1986.

Chapter 7

PLATELET STORAGE POOL DEFICIENCY

Kenneth M. Meyers and Michéle Ménard

INTRODUCTION

Platelets respond to an agonist by forming platelet agonists, such as thromboxane (TxA)$_2$ and, in some species, platelet activating factor (PAF) or by secreting platelet agonists, ADP and 5-hydroxytryptamine (serotonin, 5-HT), that are stored in dense granules.[82] Compared to 5-HT, ADP is the stronger agonist of human platelets and is, therefore, considered to be the primary stored agonist. In addition to the dense granule adenine nucleotide pool, there is a metabolic pool and a protein-bound pool.[24,46]

A reduction in the storage or dense granule adenine nucleotide pool has been termed a storage pool deficiency (SPD). SPD patients usually have a mild to moderate bleeding diathesis; however, severe life-threatening or fatal bleeding episodes have been described.[51,114,144,167]

Patients with SPD have a prolonged bleeding time with a normal platelet count. Typical platelet aggregation abnormalities include an absence of epinephrine- and ADP-induced secondary aggregation and an impaired response to low collagen concentrations, although these abnormalities are not always present.[52,98] Platelet 5-HT concentration is reduced.

A diagnosis of SPD is often confirmed by measuring platelet nucleotides. First, the nucleotides are extracted using either ethanol or perchloric acid. Ethanol extracts the dense granule and the metabolic nucleotides while perchloric acid extracts the three adenine nucleotide pools.[24] Nucleotides are then measured by high pressure liquid chromatography[24,125] or the luciferin/luciferase reaction.[47] A reduction of platelet adenine nucleotides compared to normal is assumed to reflect the SPD. The ADP content of the storage pool is greater in human platelets than is the ATP content;[49] therefore, in SPD there is an increase in the ATP/ADP ratio.[49]

The assumption that a reduction in platelet nucleotides indicates an SPD may be inappropriate if the larger metabolic pool is abnormal. Secretion[46,62,86,87,91] or controlled cell lysis[2] may be more appropriate methods of measuring the stored adenine nucleotide pool. In secretion studies, the metabolic nucleotide pool is often radiolabeled with adenine. Exchange between labeled ATP and ADP in the metabolic pool and non-labeled ATP and ADP in the storage pool is limited. After labeling, maximal secretion is induced and the nucleotide content of the medium and the specific radioactivity of the secreted nucleotides are measured. Nucleotides originating from the metabolic pool have a high specific activity, while nucleotides from the dense granule pool have a low specific activity.

Platelet SPD can be acquired or congenital. Congenital SPD can be seen as a single disease involving only the dense granule and is termed platelet storage pool disease. Some SPD patients also have a platelet alpha granule deficiency.[158] In some patients, the SPD is part of a syndrome, such as the Hermansky-Pudlak syndrome (HPS) and Chediak-Higashi syndrome (CHS). Several inherited diseases in animals are associated with a platelet SPD and are used as models for human SPD. Acquired SPDs are most commonly associated with bone marrow disorders or disorders associated with increased platelet consumption or destruction.

CONGENITAL SPD

Storage Pool Disease

Dense Granule Deficiency
Fifteen patients diagnosed as having a dense granule platelet SPD not associated with an alpha granule deficiency or other known syndromes, such as the HPS or the CHS, have been extensively studied (Table 1). Two additional studies described the general findings of 13[2] and 51[98] patients with congenital platelet SPD. Included in these studies were some patients with HPS.

Clinical signs include easy bruising, excessive bleeding from minor cuts or during surgery, persistent menorrhagia, epistaxis, and excessive bleeding after delivery.[51,114,152] Bleeding time is prolonged in the range of 7-30 minutes, and platelet counts are normal.[51,114,152,158] Akkerman *et al.*[2] found a close correlation between the dense granule ADP deficiency and the bleeding time, suggesting that platelet dense granule ADP is a primary defect in SPD. Ultrastructurally, the numbers of identifiable dense granules are reduced.[148,158]

Platelet aggregation is often abnormal, especially at lower agonist concentrations. Ultrastructurally, platelets form loose aggregates that do not remain aggregated.[148] Collagen-induced aggregation is impaired at low collagen concentrations but can be normal at high concentrations.[114,150,152] The primary aggregation phase of ADP-induced aggregation is normal, but the secondary wave of aggregation is absent most of the time.[158] Disaggregation is often seen at concentrations that usually induce a secondary wave of aggregation.[51,114,148,152] An epinephrine-induced secondary wave of aggregation is usually absent,[51,114,150,151,152] or the threshold to elicit a secondary wave of aggregation is at a higher concentration than it is for normal platelets.[151]

In two patients studied by Lages and Weiss,[63] (WA who has HPS and JD), epinephrine induced a secondary wave of aggregation. Platelets from these two patients were very sensitive to ADP, and this hypersensitivity may explain the patients' secondary aggregation wave. PAF-induced biphasic aggregation was also absent.[124] Nieuwenhuis *et al.*[98] reported on 106 patients with SPD, including 51 patients with congenital SPD. Of these 51 patients, only 7 had HPS. The frequency of normal aggregation tests was 31%. All patients who had a normal aggregation response to ADP, epinephrine and collagen had a normal aggrega-

TABLE 1
Patients with Platelet Storage Pool Deficiency (SPD)
Who Have Been Extensively Studied

Patient	Reference	Patient	Reference
SPD (n=15, 28)		JS	46 61 157
		FS	46 61 62 114
LG	46 54 61 62 148 150	LS	46 61 62 114
	151 152 155 158		
JD	62 63 150 151 158	HPS (n=31, 83)	
EP	46 61 148 150 151		
	152 155 158	EV	158
SN	152 155 158 148	MV	54 62 150 158
SK	49 152	LV	54 62 150 151 158
FA	152	SM	158
CF	46 61 62 114	MP	158
II-1	51	RZ	158
III-4(HD)	51 62	PE	150 158
IV-5 (EP)	51 62	WA	63 150 151
IV-6 (CP)	51 62	JV	151
III-8	51	NR	151
IV-12	51	RZ	62 150 151 158
LP	124	LM	151
GR	124	JA	151
13 patients	2	MR	46 61 114
		FM	62
Combined Alpha/Dense Granule		BR	114
Deficiency (n=11)		MLB	114
		MTB	114
DC	46 48 49 61 62 71	H-1	36 37
	148 149 150 151	H-2	36 37
	152 155 156 157	LD	129 132
	158 164	NG	132
RC	46 48 61 62 148 149	BB	131
	150 151 154 155	M-1	72
	156 157 158 164	II-1	2 32
SC	48 62 148 149 150	II-2	2 32
	151 154 155 156	II-3	2 32
	158 157 164	II-5	32
TC	149 151 156	II-6	32
JC	54 62 150 156 158	II-7	32
MN	158	II-9	32
ShN	158	52 patients	166
AN	158		

tion response to arachidonic acid (AA) and A23187. Israels *et al.*[52] reported that 17 patients who had a prolonged bleeding time, a decreased secretion of ATP and a decreased number of dense granules, had no demonstrable abnormalities in platelet aggregation, suggesting that SPD can be present even though platelet aggregation tests are normal.

Biochemical findings reported with SPD included a reduction in platelet ATP in 6 of 13 patients.[51,114,124,158] Platelet ADP and 5-HT were markedly reduced in all patients.[51,114,124,158] The adenine nucleotide deficiency was due to a reduction in secretable adenine nucleotides.[49,62,63,114] There was variation in the amount of ADP and ATP secreted.[2,63] For example, release was virtually absent in some patient platelets while others released between 0.1 and 0.7 μmole of ADP/10^{11} platelets compared to normal platelets which released 2.0 ± 0.17 μmoles of ADP/10^{11} platelets.[62] These findings indicate there is heterogeneity in the severity of the SPD.

Akkerman *et al.*[2] reported on an additional 13 SPD patients who had reduced platelet adenine nucleotides. Using controlled digitonin-induced cell lysis, they found that the lowered ATP and ADP content of the platelet was due primarily to a decrease in dense granule adenine nucleotides; however cytosolic ADP, but not ATP, was also reduced. There was also variation in the severity of the ADP and ATP deficiency of the dense granule nucleotide pool.

Platelet 5-HT was reduced in 11 patients reported in three studies.[51,114,158] The amount of 5-HT in platelets varied from a virtual absence (patient CF),[114] to a moderate decrease (patient EP).[49,155] The initial rate of 5-HT uptake was similar to that observed with normal platelets, but the extent of the uptake was reduced in SPD platelets.[114,155] There was considerable variation in the amount of ^{14}C-5-HT accumulated between SPD patients.[62] The uptake of ^{14}C-5-HT with time was near normal in some patients (EP and LG), while it was markedly depressed in others (SN). Other dense granule constituents such as orthophosphate[62] and calcium[61,62,63,158] were decreased. The release of calcium was variable and did not correlate with ATP release.[62] Constituents stored in platelet lysosomes[46,158] and alpha granules were present in normal amounts.[158] Platelet glycoprotein distribution was normal in two patients.[54]

The AA pathway in platelets from SPD has been studied in detail in an attempt to explain why some patients exhibit a biphasic aggregation in response to epinephrine.[63,150] In three patients described by Weiss and Lages,[150] aggregation to 0.4 mM AA was normal in two patients and slightly reduced in the third. In six patients described by Ingerman *et al.*,[51] 0.5 mM AA and PGG$_2$ induced normal aggregation. Production of malonaldehyde (MDA) in response to AA was not decreased.[150] In two patients, MDA formation following clotting and induced by maximal thrombin stimulation was normal.[124] These findings suggest that neither arachidonic metabolism nor sensitivity to AA metabolites is affected by the SPD. Epinephrine-induced MDA formation was undetectable in two SPD patients and reduced in the third.[150] On the other hand, collagen-induced MDA formation was slightly reduced at high collagen concentrations and was normal at lower concentrations in two patients and reduced in a third.[150] Since AA metabolism and response to AA metabolites was normal, these findings suggest that AA liberation was impaired in these patients and that the impairment was related to the agonist and its strength as a platelet agonist.

Combined Alpha Granule-Dense Granule Deficiency

Weiss et al.[158] described two families (family C and N) with three members each whose platelets were markedly deficient in ultrastructurally identifiable dense granules and were partially deficient in alpha granule constituents. They also described a patient (JC) whose platelets were markedly deficient in ultrastructurally identifiable alpha granules and in alpha granule constituents. Lysosomal enzymes were normal in all seven patients.

Two patients originally described by Pareti et al.[114] were subsequently shown to be deficient in heparin-neutralizing factor (HNF).[61] Patient JS, who was described by Lages et al.,[61] was also shown to be deficient in HNF. Since HNF is localized in alpha granules, these patients may also have had a combined deficiency.

Family C, originally described by Weiss et al.,[149] had a history of easy bruising, hypermenorrhagia, excessive bleeding and prolonged bleeding time from cuts, but normal platelet counts.[149,158] Collagen-induced aggregation was impaired, secondary aggregation to epinephrine and ADP was absent, disaggregation of ADP-induced aggregation was observed[149] and secondary aggregation to epinephrine was not observed, even at concentrations as high as 50 μmol/L.[151] Platelet ADP was reduced[48,149,156,158] as was release by connective tissue.[149] Holmsen and Weiss,[48,49] Holmsen et al.,[46] and Lages et al.[62] subsequently demonstrated that the defect in ADP content and release was due to a deficiency in the storage pool of ADP. The ADP deficiency was not as severe as that reported for HPS patients.[46,156] Calcium and 5-HT were reduced[49,155,156,158] but not to the extent seen in patients with HPS.[156,158] The ^{14}C-5-HT uptake was only slightly reduced in one patient.[155] Acid hydrolase content was normal, and ultrastructurally identifiable dense granules were absent in platelets from these patients.[158] Alpha granule constituents were reduced or at the lower limits of the normal range.[158] In electron micrographs, alpha granule numbers were reduced more than 50%.

The interaction of platelets from family C with the subendothelium at high shear forces was studied.[156] Platelet adhesion was normal, but thrombus volume and height were markedly reduced.[156] The size and height of the thrombus were less than that observed in HPS patients who had a more severe SPD.

Platelets from three members of family C also had depressed AA-induced aggregation, which was severe in two family members.[150] In these patients, MDA production to collagen, epinephrine and AA were markedly decreased[150] and prostaglandin production in response to thrombin was also decreased.[164] Members of this family seemed to have multiple platelet defects, including a defect in both the dense granule and AA pathways for platelet activation. These two defects may account for the pronounced reduction in thrombus formation seen in these patients.[156]

Patient JC, who is not a member of family C, had a marked alpha granule deficiency. There was a history of easy bruising, excessive bleeding from small cuts, epistaxis, and menorrhagia. Platelet counts were normal and bleeding time was > 15 minutes. Collagen-induced aggregation was markedly impaired. Epinephrine-induced secondary aggregation could not be demonstrated.[151] In studies of platelet-subendothelium interaction, adhesion at high shear force was impaired as was thrombus formation and thrombus height.

Biochemical studies of patient JC showed that platelet ADP,[150,151,156,158] calcium[158] and 5-HT[156,158] but not ATP were reduced.[156,158] Releasable ATP and ADP were also reduced.[156] The decreases were not as marked as those seen in HPS patients.[156] Ultrastructurally identifiable dense granules could not be demonstrated.[158] Lysosomal enzymes were normal. Alpha granule constituents were uniformly depressed, and the number of alpha granules was decreased.[158]

The arachidonic pathway for platelet aggregation was also impaired in JC. AA-induced platelet aggregation was absent, and PGG_2-induced aggregation was markedly reduced.[150] MDA production in response to collagen, epinephrine, and a low concentration of AA was also impaired,[150] but MDA production to high concentration of AA was normal. Thus, platelets from this patient seemed able to metabolize AA but did not aggregate to it. These platelets apparently had multiple platelet defects involving alpha granules, dense granules and the AA pathway.

Hermansky-Pudlak Syndrome (HPS)

Hermansky and Pudlak[42] originally described two patients with a triad of oculocutaneous albinism, a mild bleeding diathesis and pigmented macrophages. Although rare, the HPS has been recognized worldwide in different racial and ethnic groups. There is a high prevalence of HPS in Puerto Ricans. The oculocutaneous albinism is tyrosinase positive and the degree of cutaneous pigmentation is extremely variable.[166] All HPS patients have nystagmus, albinotic fundi and decreased visual acuity. Ceroid is present in large amounts in macrophages and also accumulates in the lungs, bladder, oral mucosal and urinary sediment. Accumulated ceroid may lead to respiratory, intestinal and urinary dysfunction. Ceroid storage in the bone marrow and urine appears to increase with age and may be absent in infants, children, and even young adults.[56,166,167,168]

The bleeding diathesis may vary from a slight bruising tendency to major hemostatic complications.[32,37,62,162] The most common signs are easy bruising, prolonged bleeding following tooth extraction, epistaxis, gingival bleeding, and extensive bleeding following delivery. The bleeding diathesis is exacerbated by acetylsalicylic acid, and two deaths have been associated with acetylsalicylic acid in HPS patients.[144,167] Bleeding times are usually prolonged beyond 20 minutes,[62,158] but shorter bleeding times have been reported.[32,37,162] The platelet count is usually normal, and abnormal primary hemostasis is associated with a platelet SPD. Platelet aggregation to collagen is absent or markedly depressed.[32,37,72,150,162] Secondary aggregation to thrombin, ADP and epinephrine (5 µmol/L) are usually absent.[32,37,150,162] However Lages, Weiss and their colleagues[62,63,151] reported on two HPS patients where epinephrine consistently induced a biphasic aggregation response. Biphasic aggregation responses to epinephrine occurred 0-40% of the time in five other patients.[151] The threshold to elicit secondary aggregation occurs at a higher epinephrine concentration in HPS patients.[151]

The dense granule SPD is severe with HPS. Platelet 5-HT is markedly reduced[32,37,63,114,155,158,162] with 5-HT levels about 10% of normal.[37,63,114,155,158,162] Platelet ATP is below the normal range in about half the patients.[32,37,63,114,158] On the other hand, ADP is markedly reduced, usually less than 1 μmole/10^{11} platelets.[37,63,114,158]

Secretion and digitonin-induced platelet lysis studies have been performed, and in these studies the dense granule nucleotide pool was markedly deficient. Lages *et al.*[62] reported on 17 patients with SPD, 4 of whom had HPS. Secretable ATP and ADP were reduced, but retained constituents were normal in the four HPS patients. Akkerman *et al.*,[2] using controlled digitonin-induced platelet lysis, measured the ATP and ADP content in the cytosolic and dense granule pool of three HPS patients. These patients had unmeasurable levels of granule ATP, and granule ADP levels were less than 0.2 μmoles/10^{11} platelets. Platelet levels of other dense granule constituents, such as pyrophosphate[62] and calcium,[62,158] were reduced as was the amount that could be secreted. Calcium release was induced by thrombin even though ATP was not secreted.[62]

Lysosomal and alpha granule constituents appear to be present in normal amounts in HPS since platelet acid phosphatase, β-galactosidase, β-N-acetyl glucosaminidase, and β-glucuronidase contents are not decreased[37,158] nor are the alpha granule constituents, platelet factor 4, β-thromboglobulin, platelet derived growth factor and fibrinogen.[158] Secretion of β-hexosaminidase, and β-glucuronidase is decrease in response to low, (0.04 U/ml) thrombin concentrations, but normal at high (5 U/ml) thrombin concentrations and at 4 and 12 μM of A23187.[60] The secretion defect is corrected by ADP addition. An HPS patient with a secretion defect of acid hydrolases and alpha granule constituents was reported by Rendu *et al.*,[131] and addition of ADP partially corrected the secretion defect. Thrombin-induced phosphorylation of proteins with apparent molecular weights of 43,000 Da and 20,000 Da were similar in HPS patients to that observed in normal platelets.[131] Membrane glycoproteins were also reported to be normal in two HPS patients.[54]

TXA$_2$ formation, as indicated by TXB$_2$ or MDA production following platelet activation with AA and PGG$_2$, is normal in HPS patients.[72,131,150] On the other hand, aggregation responses to AA and PGG$_2$ are often reduced[72,150] but can also be normal.[131] TXB$_2$ and MDA formation induced by epinephrine,[63,150] collagen,[72,131,150] and thrombin[131] is decreased while A23187-induced TXB$_2$ formation is normal.[131] ADP induces a modest TXB$_2$ formation in both control and HPS platelets.[131] When ADP and thrombin are added in combination, there is a marked increase in TXB$_2$ formation to levels approaching or exceeding TXB$_2$ formation by controls.[131]

The most consistent diagnostic feature of the HPS is the lack of platelet dense granules.[166] Maurer *et al.*[77] described patients with a clinical syndrome similar to HPS where dense granules could not be observed by electron microscopy. White *et al.*[162] described a HPS patient whose platelets had 1 dense body in 20 platelets. The virtual absence of dense granules has been confirmed using platelet whole mounts,[166] standard electron microscopy,[161,165] and the uranaffin reaction.[133] After mepacrine loading, the number of mepacrine granules in platelets from two HPS patients was reduced from a mean of 5.4 granules/platelet to 1.7 and 2.9/platelet.[132] Platelet alpha granules and mitochondria were present in normal numbers and appeared ultrastructurally normal.[158,161] Recently, Gerrard *et al.*[30] found that a 40 kDa dense granule membrane protein was markedly deficient in one HPS patient.

HPS Variant

Rendu and co-workers[129] described a female patient with albinism, nystagmus and a hemorrhagic tendency. Platelet dense granule and constituents were markedly reduced. This patient had some clinical features not typical of the HPS, including a leukopenia, unusual skeletal malformations, splenomegaly, and enlargement of those dense granules that were present. Platelet phospholipase A_2 activity was undetectable.

Suggested Animal Models of the HPS

Pale ear pigment mutant mice (C57BL6J ep/ep) have been suggested as animal models for the HPS by Novak *et al.*[101] These mice have pale ears, prolonged bleeding times, decreased levels of platelet 5-HT, ATP and ADP, and marked reduction of releasable ATP and ADP. While 5-HT content is reduced, it is not reduced to the extent seen in beige (bg/bg) mice, which have been proposed as an animal model for the CHS[70] and where there is a virtual absence of dense granules. The number of dense body platelets in air-dried whole platelet mounts is markedly decreased, with a mean of 1.2 dense bodies/platelet in ep/ep mice compared to 6.4 dense bodies/platelet in control mice. Pale eared mice clearly have a platelet SPD. The platelet contents of the lysosomal enzymes, β-glucuronidase and β-galactosidase are normal in ep/ep platelets. The pigmentary dilution may be due to abnormally small melanosomes.[119] Serum levels of acid hydrolases are increased and secretion of kidney lysosomal enzymes is defective.[104] These alterations, to our knowledge, have not been described in HPS patients, but abnormal kidney lysosome function has been suggested in HPS patients.[168] On the other hand, ceroid storage has not been reported in the ep/ep mouse. These apparent differences in platelet 5-HT and lysosomes between the HPS and the ep/ep mouse suggest that, while the ep/ep mouse is an important animal model of SPD, it is not clear if it is an animal model of HPS. Ceroid accumulation has been described in the light-ear mutant mice (le/le),[79] but the platelet 5-HT content of these mice is considerably higher than that of bg/bg mice,[102] suggesting that the SPD is not as severe as reported

TABLE 2
Platelet serotonin content of mice with a SPD

Genotype	5-HT (μg/10⁹ platelets)	Ref.
C57BL/6J Normal	5.29±0.51	101
Beige (bg/bg)	0.04±0.01	102
Pale ear (ep/ep)	1.05±0.16	101
Light ear (le/le)	1.12±0.12	102
Pallid (pa/pa)	0.04±0.01	102
Maroon (ru-2mr/ru-2mr)	0.32±0.06	102
Ruby-eye (ru/ru)	0.33±0.05	102
CoCo (coa/coa)	0.40±0.10	105
Pearl (pe/pe)	0.14±0.09	102

in the HPS. Age related ceroid deposition has also been described in the bg/bg mouse,[107] but the enlarged granules in the bg/bg mouse suggests that it is a model of the CHS.[70]

A platelet SPD has also been described in pearl, maroon, ruby-eye, pallid, and coco mice.[102,105] The levels of 5-HT vary between the mutants (Table 2). The mutants with the lowest 5-HT levels are the beige and pallid mice while the pale ear and light ear mice have the highest amounts of 5-HT.

There are eight distinctly different genetic mutations in mice that result in the phenotypic expression of a platelet SPD. There is variation in the severity of the SPD, clearly showing the heterogenity of platelet SPD.

Chediak-Higashi Syndrome (CHS)

The CHS is an autosomal recessive disorder with enlarged granules.[10,16,43,141] Mink[65] with a similar syndrome were first described at Washington State University (WSU). Subsequently, the syndrome has been described in cattle,[110] mice,[69] cats,[58] foxes,[96,97,138] rats[100] and a killer whale.[143] The CHS is a triad of partial oculocutaneous albinism, recurrent pyogenic infections, and a bleeding tendency. Enlarged granules, including lysosomes, are present in most granule forming cells.[70] The most consistent diagnostic pathological finding is the presence of giant granules in peripheral blood leukocytes.[12,106,160] Cytoskeletal abnormalities,[123] impaired cyclic nucleotide metabolism,[14] defective protein kinase C activity,[53,137] leakage of hydrolytic enzymes from cytoplasmic organelles,[56] and defective cellular organelle membranes[35] have also been described.

A hemorrhagic tendency in CHS cattle and mink was noted by Padgett[109] in a review of the CHS. A detailed study of the coagulation and fibrinolytic systems of CHS cattle and mink did not disclose a cause for the hemorrhagic tendency.[118] Platelet counts were normal or elevated.[86] Page et al.,[111] in an attempt to explain the pigmentary dilution in CHS, measured whole blood 5-HT while studying tryptophan metabolism in two children with the CHS. He reported a consistent lack of measurable 5-HT in both patients while their non-affected parents had normal levels. Holland[44] found that bleeding times in CHS mice were prolonged and that the concentration of 5-HT/5-HT metabolites in whole blood from CHS cattle, mink, and mice were significantly lower than in control animals. These findings were confirmed in CHS mice and cattle by Meyers *et al.*[93] who also reported that platelet 5-HT uptake was decreased, as previously observed in SPD human platelets. In CHS mice, the reduction in 5-HT/5-HT metabolites was found to result from a decrease in platelet 5-HT. Brain 5-HT levels and synaptosomal uptake of 5-HT were normal.[83] Aggregation of CHS bovine platelets to ADP was subsequently reported to be normal, but collagen-induced aggregation was depressed and platelet adenine nucleotide concentration decreased.[8]

A series of articles by Bell *et al.*,[9] Buchanan and Handin[15] and Costa *et al.*[18] followed, reporting that human CHS patients were deficient in platelet adenine nucleotides, especially ADP, and the platelet ATP/ADP ratio was increased. Boxer *et al.*[13] reported that secretable adenine nucleotides were virtually absent in an infant with CHS. Platelet concentration and secretion of ATP and ADP was reduced in one CHS patient[116] and in six other CHS patients.[4]

A marked reduction in platelet calcium[4,13] and 5-HT[4,9,13,18] is present in CHS patients. Initial uptake of 5-HT by human CHS platelets is normal or slightly reduced, but total uptake is markedly reduced[4,13,15,116,128] and is accompanied by increased 5-HT metabolism.[128] The reduction in dense granule constituents to levels approaching those reported for the HPS suggest that most CHS patients have a severe platelet SPD (Table 3).

Constituents stored in other granules are not altered. Acid hydrolases content was found to be normal in platelets from 10 CHS patients.[4,13,128] Beta-thromboglobulin and platelet factor 4 levels, normally stored in alpha granules, were normal in three CHS patients.[128] Aggregation responses of platelets from CHS patients are presented in Table 4.

Platelets from CHS patients have a reduced number of platelet dense bodies. In air-dried whole mounts, platelets from two CHS patients had only 10% of the normal dense body numbers.[18] Parmley *et al.*[116] and Rendu *et al.*[128] also noted, in electron micrographs of platelets fixed with glutaraldehyde in cacodylate[116] or in White's saline,[116] that the number of dense granules was reduced. CHS platelets from the three patients had 5%, 10% and 20% of the dense bodies seen in thin sections of normal platelets, although a few platelets had an excessive number of dense bodies.[128] Rendu *et al.*[128] labeled platelets from three CHS patients with mepacrine to visualize dense granules. They reported that the fluorescent granules were reduced in number and were smaller or less fluorescent than those seen in normal platelets. The mean numbers of fluorescent granule platelets were 3.4, 3.0 and 1.8 for the three patients studied, while control values were 5.5 + 0.8. There appears to be heterogenity in SPD of CHS patients. White[161] observed a CHS patient without a dense granule SPD for over 15 years. Most patients with CHS have a severe platelet dense granule SPD.

TABLE 3
Platelet content of dense granule constituents in CHS patients

Number of Patients	5-HT*	ATP*	ADP*	Ca*	Ref.
1		2.4	1.03		116
2		5.5	0.45		15
		5.8	0.85		
1	0.13	2.12	0.36	5.45	13
2	0.04	2.34	0.88		9
	0.03	2.48	0.91		
6	0.15	1.65	0.3	4.66	4

* Data expressed as μMoles/10^{11} platelets

TABLE 4
CHS Platelet Aggregation Responses

Agonist	Response	Number of patients	Ref.
ADP	normal extent deaggregation	6	4
	essentially normal	2	9
	normal	1	116
		3	128
		2	15
Collagen	impaired	6	4
		2	9
		2	15
	normal(very high concentration)	1	116
Epinephrine	secondary aggregation absent in 4 and reduced in 2; did not respond to 5 lM	6	4
	secondary wave absent	2	9
		1	116
		2	15
A23187	normal in 2 and abnormal in 4	6	4
Thrombin	greatly reduced	3	128

It has been suggested that an abnormality in CHS leukocytes in cyclic nucleotide metabolism prevents microtubule polymerization.[108] In thin sections, the microtubule coil in CHS platelets appears normal.[116,128,160] In response to collagen, CHS platelets change shape and granules become centralized, encircled by microtubules as observed with normal platelets.[160] This process and the temperature-induced disassembly and then reassembly of CHS platelet microtubules was not altered in CHS platelets.[160] These findings indicate that the microtubule defect of CHS leukocytes does not extend to platelets. A CHS patient was shown to have elevated blood cAMP levels and abnormal polymorphonuclear leukocyte function, defects that were successfully treated by ascorbic acid.[13] Ascorbic acid treatment did not correct the platelet SPD in this patient [13] even though cAMP levels fell and remained low after ascorbate was discontinued, suggesting that the SPD in the CHS is not associated with a cAMP alteration.[4,13]

Enlarged lysosomes have been reported in CHS platelets.[116,128,160] White[160] reported that giant granules could be seen in platelets from a CHS patient but the frequency was approximately 1/1000 platelets in thin sections. A higher frequency of giant granules in another patient was reported by Parmley *et al.*[116] Approximately 5% of the platelets in thin section contained giant granules. Parmley *et al.*[116] also noted giant granules within CHS megakaryocytes (MK) from one patient. Alpha granules[116,128,160] were normal, as were microperoxisomes.[128]

CHS Cattle

Hereford cattle affected with the CHS were first described at WSU. Recently the CHS has been identified in Brangus Cattle at Kansas State University.[5] The cattle at WSU had markedly prolonged bleeding times[91] with normal coagulation tests.[118] Platelet numbers were usually elevated, being 175% that of control cattle.[86] This elevation in platelet count was not observed in other species with the CHS, indicating that it is a species difference rather that a feature of the syndrome. Cattle with the CHS generally have chronic pneumonia, and numerous abscesses are commonly observed on post mortem. The possible relationship between chronic inflammation and thrombopoiesis awaits clarification.

Resting and Secreted Granule Constituents

Platelet ATP and ADP are reduced[86,91] with negligible release by thrombin stimulation.[86] The ATP and ADP in the suspension medium following thrombin stimulation has a high specific activity and is similar to unstimulated CHS platelets, suggesting that the storage pool of adenine nucleotides is completely absent in these platelets.[86]

Unlike in CHS cats and mink, 5-HT can be measured in CHS bovine platelets. This may reflect the higher 5-HT content of bovine platelets compared to feline or mink platelets.[85] CHS cattle platelets contained only about 10% of the 5-HT found in normal cattle platelets. Using 4,6-difluoro-5-HT and nuclear

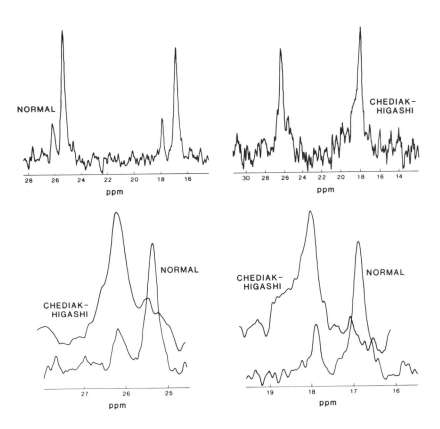

FIGURE 1. The ^{19}F NMR spectra from normal cattle and cattle with the CHS. Platelets were incubated with 4,6-difluoro-5-HT for 1 hr at 37°C, washed and ^{19}F NMR spectra obtained at 188 MHz.[84] (Reprinted with permission from *Thromb. Res.* 58, Copyright 1990, Pergamon Press, Inc.)

magnetic resonance, the dense granule 5-HT pool was found to be below detectable limits in CHS cattle platelets (Figure 1).[84]

Total platelet magnesium and, to a lesser degree, platelet calcium were decreased, and the amount that could be secreted was also reduced.[86] The percentage of the total Ca^{2+} released by bovine CHS platelets was similar to that of normal platelets. On the other hand, the percentage of Mg^{2+} that was maximally released was about half that of normal platelets. The release of Ca^{2+} (Figure 2) and Mg^{2+} (Figure 3) depended upon the dosage of thrombin used. The kinetics of thrombin-induced release of Ca^{2+} and Mg^{2+} from normal and CHS platelets was similar (Figure 4). Approximately 70% of the release was completed by 30 sec. While the intracellular location of the releasable metals in bovine platelets has not yet been identified, the dose response and secretion rate suggest a granule location.

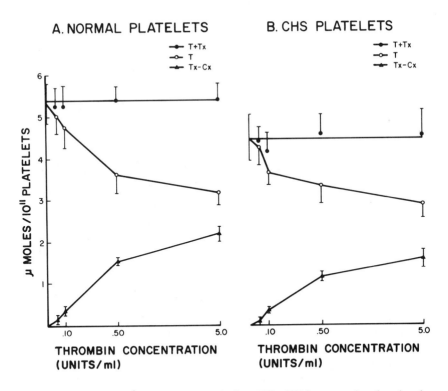

FIGURE 2. Release of Ca^{2+} from gel filtered platelets (GFP). GFP from normal cattle and cattle with the CHS were treated with increasing concentrations of thrombin or 0.15 M NaCl as described previously.[86] Three minutes later the platelets were centrifuged and the Ca^{2+} concentration of the platelets or the suspension medium determined using an atomic absorption spectrometer. T refers to the Ca^{2+} within the cell, Tx refers to Ca^{2+} in the suspension medium, and Cx refers to Ca^{2+} in the suspension medium of GFP treated with the vehicle control. Release was calculated as Tx-Cx.

Ultrastructure

Platelets from CHS cattle appeared normal with transmission electron microscopy, except they lacked dense granules.[80,86,120] As in CHS cats and mink, giant granules were not observed with routine fixation[86] in CHS cattle platelets. Lysosomes could be demonstrated in thin sections using acid phosphate cytochemistry and cerium as a trapping agent.[81] Primary and secondary lysosomes were observed in normal and CHS platelets in equal numbers. Lysosomes in CHS platelets were neither enlarged nor was there a difference in the primary to secondary lysosome ratio. Lysosomal enzymes acid phosphatase and β-N-acetylglucosaminidase were found in normal amounts and β-N-acetylglucosaminidase release was normal.[91] It is not clear why platelet lysosomes are not enlarged in CHS cattle since lysosomes are enlarged in most other cell types.

Platelets from CHS cattle were processed for electron microscopy after stimulation with 10 μM ADP, a concentration that induces maximal aggrega-

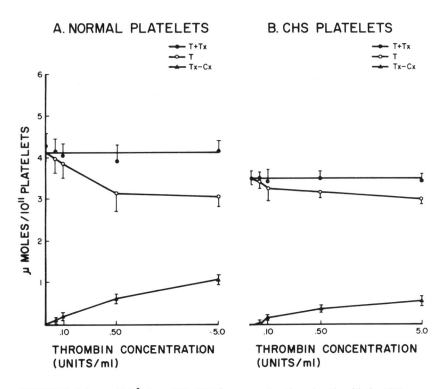

FIGURE 3. Release of Mg^{2+} from GFP. GFP from normal cattle and cattle with the CHS were treated with increasing concentrations of thrombin or 0.15 M NaCl as described previously.[86] Three minutes later the platelets were centrifuged and the Mg^{2+} concentration of the platelets or the suspension medium determined using an atomic absorption spectrometer. T refers to the Mg^{2+} within the cell, Tx refers to Mg^{2+} in the suspension medium, and Cx refers to Mg^{2+} in the suspension medium of GFP treated with the vehicle control. Release was calculated as Tx-Cx.

tion of normal cattle platelets (Figure 5).[88] During platelet shape change, platelets lost their discoid shape and the granules moved toward the periphery. Only small platelet aggregates formed. When aggregation was about 50% of maximum, large platelet aggregates were seen. Areas in the aggregates with activated platelets in very close contact and numerous pseudopod projections were less frequent with CHS platelets compared to normal platelets in these areas and were less activated. After six minutes, when aggregation was still maximal, two types of aggregates could be demonstrated with normal platelets. Some aggregates contained degranulated platelets in the center with platelets containing granules in the periphery projecting pseudopods toward the center. Other aggregates (Figure 6) had a core composed of platelet containing granules in the center and degranulated platelets in the periphery. The latter type was not seen with CHS platelets and CHS platelets appeared less activated (Figure 5D). These findings suggest that bovine CHS platelets can respond to an agonist like normal bovine platelets, although the response is weaker.

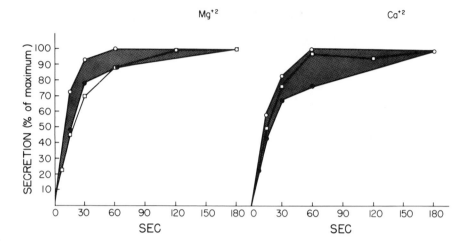

FIGURE 4. Release of Ca^{2+} and Mg^{2+} into the suspension medium from CHS GFP treated with 5 U/ml of thrombin as a percent of maximal release. GFP was treated with thrombin or 0.15 M NaCl. At the times indicated, platelets were combined with ice-cold 135 mM formaldehyde: 8.9 mM ethylenediaminetetracetic acid, the mixture centrifuged and Ca^{2+} and Mg^{2+} in the supernatant measured by atomic absorption spectroscopy. Release was calculated as Tx-Cx.

Basis for the SPD

The virtual absence of dense granule ATP, ADP, 5-HT, and identifiable dense granules suggests that the dense granule or its precursor or ghost is absent in CHS cattle platelets. This could result from a lack of formation of the granule by MK or from the rapid fusion of these granules. Alternatively, the SPD could result from an abnormality where dense granule ghosts are present but granule membrane processes necessary to adequately accumulate or store dense granule constituents are lacking. To test these alternative hypotheses the following questions were asked:

(1) Can functional processes considered unique to the dense granule be demonstrated in CHS platelets?

(2) Can proteins considered specific for dense granule membranes be found in CHS platelets?

(3) Can dense granules be identified using special staining procedures?

(1) *Can functional processes considered unique to the dense granule be demonstrated in CHS platelets?* When platelets are shed from MK, dense granule precursors contain nucleotides.[80] The 5-HT is accumulated by platelets while they circulate,[38] and it enters the platelet by an imipramine sensitive transporter.[90] Inside the platelet, 5-HT can be conjugated,[23] metabolized to 5-hydroxyindolacetaldehyde and then to either 5-hydroxyindoleacetic acid (5HIAA) or 5-hydroxytryptophol (5-HTOH),[7] enter the dense granule using a reserpine sensitive transporter that is driven by a H^+ gradient,[134] or leave the platelet.

CHS cattle platelets accumulate and maintain ^{14}C-5-HT for short periods of time (hours) but cannot store 5-HT. Maintenance of ^{14}C-5-HT within CHS platelets is due in part to the plasma membrane imipramine sensitive transporter.[90] There is enhanced metabolism of ^{14}C-5-HT to ^{14}C-5-HTOH by CHS platelets, although enhanced metabolism of 5-HT cannot account for the reduced 5-HT levels. Some ^{14}C-5-HT is being taken up by granules in CHS platelets as evidenced by the rapid release following thrombin stimulation.[90]

To identify the subcellular location of the ^{14}C-5-HT, platelets were disrupted and the granules subfractionated using sucrose gradient ultracentrifugation.[90] The mixed granule fraction, obtained after disruption, separated into three types of granules after density gradient ultracentrifugation (Figure 7). The mitochondrial fraction was located at the junction between 1.2 M and 1.4 M sucrose, alpha granules were found between 1.6 M and 1.8 M sucrose with a heavy concentration at the junction between 1.6 M and 1.8 M sucrose, and the dense granule pellet was below 2.0 M sucrose. The CHS granules containing 5-HT were located in the 1.6 M sucrose solution, or the lighter alpha granule fraction. Acid hydrolases, presumably lysosomes, were found in this fraction when human platelets were similarly processed.[29]

The ^{14}C-5-HT was taken up by isolated mixed granules from CHS platelets.[90] Uptake of 5-HT was dependent upon an acidic intra-granule environment, and uptake could be reversed by nigericin or NH_4Cl, agents that dissipate granule proton gradients. Unlike 5-HT uptake by normal platelet dense granules, uptake by CHS mixed platelet granules was not inhibited by reserpine. If dense granules were present in CHS platelets, but could not generate a H^+ gradient, they would not accumulate 5-HT. In studies done in collaboration with Gary Dean, CHS mixed granules were acidified by incubating the granules in K^+ and, after washing, nigericin was added to exchange internal K^+ with external H^+. This was done to determine if CHS platelets contained abnormal dense granules which were unable to generate a H^+ gradient, and thus unable to accumulate 5-HT. These acidified mixed granules did not take up 5-HT by a reserpine sensitive process. Results from these studies indicate that 5-HT is not being accumulated by dense granules even though 5-HT accumulates in an acidic granule where it becomes ionically trapped. It is likely that 5-HT accumulates within lysosomes since these granules are acidic and 5-HT is found in the lysosome-enriched granule subfraction.

(2) *Can proteins considered specific for dense granule membranes be found in CHS platelets ?*[89] Platelet granules were obtained by ultracentrifugation after platelet disruption using a French pressure cell. The mixed platelet granules were subfractionated into mitochondrial-, alpha granule- and dense granule-fractions by sucrose gradient ultracentrifugation. Surface proteins of the intact granules were radiolabeled and the granules lysed. Proteins in the granule membranes were extracted and then separated by sodium dodecyl sulfate (SDS) gel-electrophoresis. All radiolabeled proteins present in the

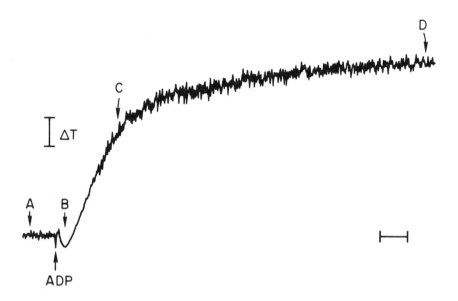

FIGURE 5. Ultrastructure of ADP-induced platelet aggregation. PRP from normal and CHS cattle were treated with 10 μM ADP. Samples were removed indicated by the arrows on the aggregation tracing. Samples B, C, and D are presented.[88] Horizontal bar denotes 0.5 minutes. (Reprinted with permission from *Am. J., Pathol.* 106, Copyright 1982, J.B. Lippincott Co.)

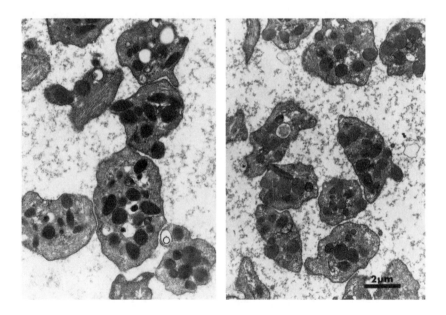

FIGURE 5B. Aggregates of platelets from normal cattle (left) and CHS cattle (right) sampled at time B.

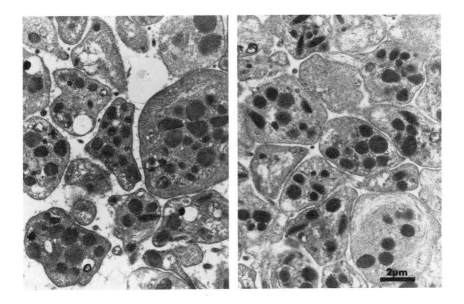

FIGURE 5C. Aggregates of platelets from normal cattle (left) and CHS cattle (right) sampled at time C.

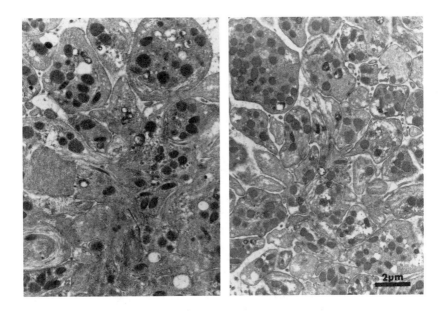

FIGURE 5D. Aggregates of platelets from normal cattle (right) and CHS cattle (left) sampled at time D.

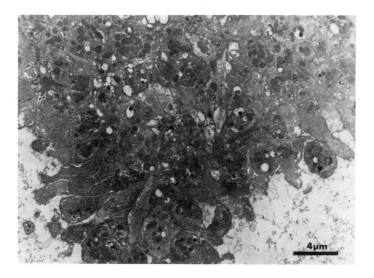

FIGURE 6. Electron micrograph of a platelet aggregate obtained from normal cattle PRP that had been fixed 6 minutes after adding 5 μM ADP.

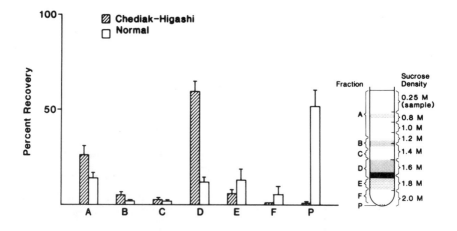

FIGURE 7. Subcellular distribution of 5-[^{14}C]HT in a sucrose density gradient. Platelets from normal cattle and CHS cattle were incubated with 5-[^{14}C]HT and then disrupted using a French pressure cell. The mixed granule fraction was placed onto a sucrose density gradient and ultracentrifuged. The right diagram shows sucrose density gradient and subfractions(A-P) into which it was divided. Fraction A consisted of membranes, B of mitochondria, D and E of alpha granules and F of dense granules.[90] (Reprinted with permission from *Am. J. Physiol.* 245, copyright 1983, American Physiological Society.)

mixed granule fraction could be identified in the three granule subfractions. In the dense granule subfraction there were three prominent proteins (Figure 8). The dense granule proteins could not be identified in the mixed granule fraction nor in any granule subfraction of CHS platelets (Figure 8). These findings also suggest that functionally impaired dense granules are not present in CHS platelets. Furthermore, these studies suggest that fusion of dense granules or its precursor with other granules cannot account for the platelet SPD.

(3) *Can dense granules be identified using special staining procedures?* Cattle platelets do not have a membrane invagination system, i.e., they lack an open or surface-connected canalicular system.[171] Therefore, internal structures should be organelles. Clear vesicles bounded by a sharp membrane, which have been shown to be dense granule precursors,[80] were not identified in CHS platelets fixed with glutaraldehyde in either cacodylate (Menard, personal

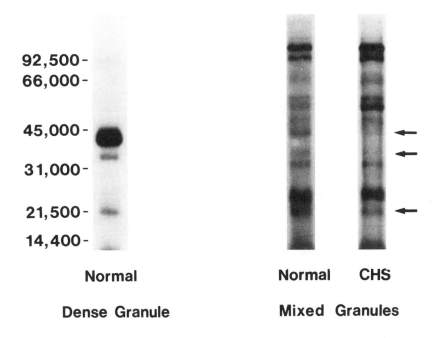

FIGURE 8. Autoradiograms of one-dimensional electrophoretograms of mixed granule membrane protein from normal cattle and CHS cattle platelets and of dense granule membrane protein from normal cattle platelets. Platelets were disrupted with a French pressure cell and the mixed granule fraction obtained by differential centrifugation. Dense granule-enriched subfractions were obtained by discontinuous sucrose gradient ultracentrifugation. The surface proteins of the mixed granule fraction and the dense granule-enriched subfraction were radiolabeled with ^{125}I using lactoperoxidase and the granule osmotically lysed. The membrane proteins were extracted with Triton X-100 and electrophoreses after reduction in a 7% SDS-polyactylamide gel.[84] (Reprinted with permission from *Thromb. Haemost.* 64, copyright 1990.)

observation) or a calcium-enriched buffer.[88] In normal platelets, dense granules can be readily identified with the uranaffin reaction. Fixed platelets are incubated with uranyl acetate and dense granules appear as a vesicle which has an eccentrically located electron-dense core and electron-dense material associated with the interior side of the granule membrane. The electron-dense precipitate is thought to result from a reaction between uranyl ions and the phosphate groups of phosphonucleotides.[68] This structure is not present in CHS MK[80] nor in platelets (Menard, personal communication).

Normal and CHS platelets processed with standard fixation protocols have membrane bounded structures (MBS) with amorphous material of varying densities (Figure 9).[88] When bovine platelets are stained for acid hydrolases, these structures are replaced with phosphatase-positive granules,[81] suggesting that the MBS are lysosomes. Structures others than previously described granules are not seen in glutaraldehyde fixed CHS platelets.

There is no ultrastructural nor biochemical evidence for the presence of normal or abnormal platelet dense granules in CHS platelets. There are granules

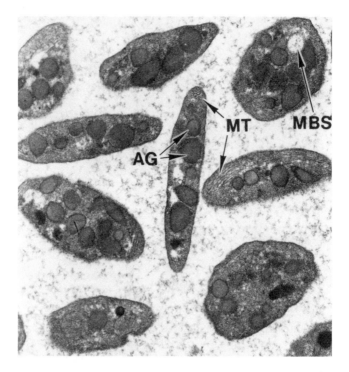

FIGURE 9. Electron micrograph of CHS PRP. AG, alpha granule; MT, microtubules; and MBS, membrane bounded structure.[86] (Reprinted with permission from *Am. J. Physiol.* 237, copyright 1979, American Physiological Society.)

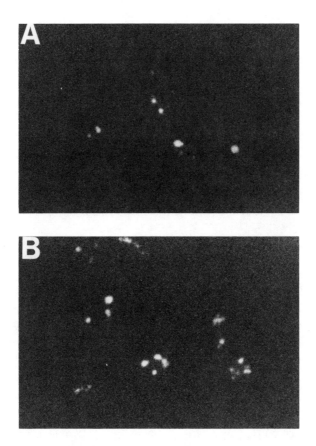

FIGURE 10. Mepacrine containing granules in platelets from normal cattle (A) and in cattle with the CHS (B). Platelets were incubated with 0.5 µM mepacrine, after adjusting the platelet count to 5,000,000 platelets/µl, for 30 minutes at 37°C and isolated from plasma by gel filtration. The gel-filtered platelets (GFP) were then processed as described by Lorez *et al.*[67] For quantitative expression, mepacrine containing granules were counted by noting the number of granules within platelets until the granules began to flash or fade. Generally, only 1-2 platelets could be counted before it was necessary to move to a fresh field. Two hundred platelets from three normal cows and three-hundred platelets from three CHS cows were counted. One normal cow and one CHS cow were processed simultaneously.

that can accumulate but not store 5-HT. These granules do not have character-istics of dense granules and are most likely lysosomes.

Mepacrine Containing Granules in CHS Cattle Platelets

Mepacrine is a basic tricyclic amine with fluorescent properties. In cells other than platelets, mepacrine accumulates in lysosomes,[3] and in platelets it

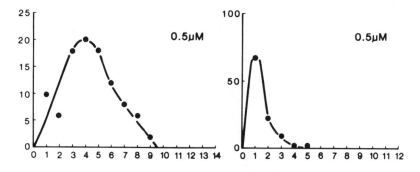

FIGURE 11. The frequency distribution of mepacrine containing granules in normal cattle platelets and CHS cattle platelets processed as described in Figure 10. The abscissa is the number of mepacrine containing granules per platelet and the ordinate is the percent of the platelet population.

accumulates in dense granules.[139] Normal cattle platelets contain an average of five mepacrine granules/platelet (Figure 10). CHS platelets have an average of two mepacrine-containing granules (Figure 11). Since there is no physiological, biochemical or ultrastructural evidence that CHS cattle platelets have dense granules in any form, it is likely that mepacrine is present in another granule in CHS platelets. We have shown that platelets contain lysosomes and that there are acidic granules that can accumulate 5-HT. These findings suggest that in the absence of dense granules, mepacrine will accumulate in lysosomes as it does in other cells that lack dense granules. It, therefore, appears unlikely that mepacrine can be used as a marker for platelet dense granules in SPD platelets.

Beige Mice (bg/bg)

Beige mice had an increased bleeding time (>10 minutes) with normal platelet numbers.[45,102] Platelet ATP,[45,102,133] ADP[45,102] and 5-HT[45,102,103,133] were markedly decreased. The initial uptake of ^{14}C-5-HT was not affected, although the extent of uptake was decreased.[102] Injecting 5-HT intravenously was reported to correct the prolonged bleeding time.[45] Levels of acid hydrolases were normal, but thrombin-induced secretion of these enzymes was decreased.[102] Platelet dense granules were not seen in thin sections from glutaraldehyde fixed platelets in phosphate buffer[45] nor in White's buffer.[127] However, electron-lucent cores surrounded by a clear halo were occasionally seen.[45] Platelets and MK from bg/bg mice were reported to give neither a uranaffin nor a chromaffin reaction.[133]

Mepacrine with its strong affinity for platelet dense granules, has also been used to characterize the platelet dense granule deficiency. In spite of the virtual absence of 5-HT in bg/bg platelets, the number of mepacrine-containing granules was reduced by only 35% in bg/bg mice.[142] The fluorescent characteristics of the granules differed from that of normal platelet granules, the fluorescent intensity being less than in normal platelets[142] and the number of flashes being markedly decreased.[142] In contrast, Reddington et al.[127] reported that bg/bg mice had a normal number of mepacrine-containing granules/platelets. As with studies by Lorez and Da Prada,[67] the number of flashing granules and the relative fluorescent intensity was markedly decreased.[127] The SPD in bg/bg mice is a defect in the pluripotential progenitor bone marrow cell, since bone marrow transplantation corrects the bleeding time and restores platelet 5-HT.[103]

Rats, Foxes, Mink and Cats with the CHS

The CHS has recently been described in the DA strain of rat, designated bg/bg.[100] These rats have prolonged bleeding times and marked reductions of platelet 5-HT concentrations. Administration of 5-HT corrects the bleeding time.[100]

The CHS has been described in blue foxes[96] and silver foxes.[97] Sjaastad et al.[138] studied the platelet defect in 18 blue foxes with the CHS. Bleeding times were greatly prolonged. Platelet counts were normal, but aggregation to ADP, collagen, and 5-HT was impaired. Platelet 5-HT was extremely low, and ATP and especially ADP were reduced. A CHS silver fox was shown to lack osmiophilic platelet dense granules, and about 13% of its platelets had enlarged granules.[27] AA-induced aggregation was impaired in these animals.[138]

Mink[87] with the CHS had prolonged bleeding times with normal platelet counts and coagulation times. Collagen-induced aggregation was impaired, especially at low collagen concentrations, and 5-HT was below levels of detection.[87] Platelet ATP and ADP were reduced and the ATP/ADP ratio was increased.[87] Secretable stores of ATP and ADP were virtually absent upon stimulation of gel-filtered platelets (GFP) with thrombin.[87] The specific activity of the released nucleotides was very high and similar to the specific activity of GFP from CHS mink, suggesting that the released nucleotides originate from the metabolic cytoplasmic pool and not from dense granules. Thrombin-induced release of both calcium and Mg was impaired.[87] Dense granules were not seen in thin sections of CHS mink platelets.[88] The AA pathway for platelet activation appeared normal since MDA production by CHS platelets in response to 0.5 and 1.0 mM of AA was similar to control platelets.[87] MDA production in response to thrombin was slightly increased.[87]

A colony of cats affected with CHS is kept at Washington State University. These cats have prolonged bleeding times with normal platelet counts.[91] Platelet aggregation to ADP, 5-HT and collagen is impaired (Figure 12). Platelet ATP and especially ADP are markedly reduced. Secretable ADP is below the level of detection. Secretable ATP is very low and the specific activity is very high, suggesting that ATP originates from the cytoplasmic ATP pool. Secretable Ca^{2+} and Mg^{2+} are decreased. Platelet 5-HT is virtually absent. The lysosomal

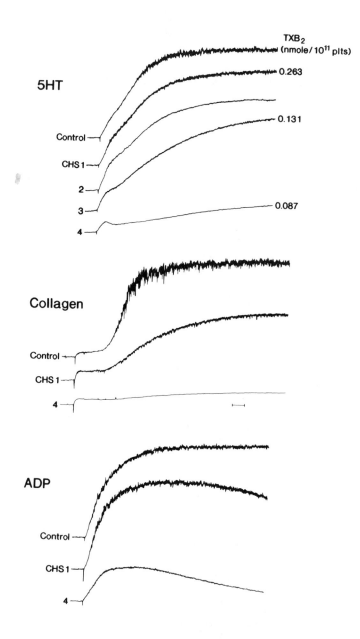

FIGURE 12. Platelet aggregation response of platelets from four CHS cats and one normal cat to 5-HT (12, 5 μM), collagen (25 μg/ml), and ADP (5 μM). The amount of TxB$_2$ formed by platelets from three CHS cats in response to 12.5 μM 5-HT is presented.[91]

constituents, acid phosphatase and β-N-acetylglucosaminidase, are present in normal amounts. Platelet dense granules are not seen with transmission electron microscopy.[88] CHS feline platelets appear to have a normal AA pathway for platelet aggregation. The extent of platelet aggregation, aggregation slope, and whether there is biphasic aggregation depends upon the amount of TxB_2 formed by CHS cat platelets in response to an agonist (Figure 13). AA induces irreversible aggregation of CHS cat platelets,[91,92] and CHS cat platelets form normal amounts of TxB_2 and MDA in response to thrombin or AA.[91]

There are numerous cell types affected with the CHS, and multiple defects have been reported for the CHS. It, therefore, cannot be assumed that the prolonged bleeding seen in the CHS is due entirely to platelet SPD. In a recent study by Cowles *et al.*,[22] CHS cats were transfused with platelet concentrates from normal cats with minimal erythrocyte and leukocyte contamination. Bleeding times normalized following the transfusion, suggesting that the prolongation of the bleeding time in CHS cats can be attributed to a platelet defect. Recently, whole blood aggregation tracings from heterozygous CHS cats were reported to be abnormal, suggesting that there may be another platelet defect besides the SPD in the CHS cat line.[17]

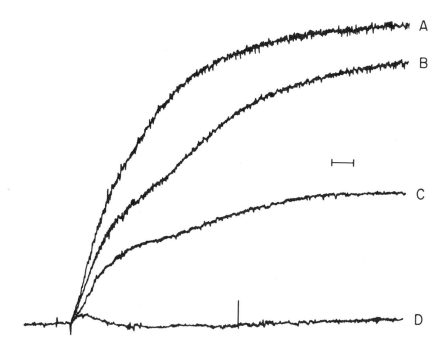

FIGURE 13. The 5-HT-induced aggregation of platelets from CHS cat 1 in Figure 12. A = 25 μM, B = 12.5 μM, C = 6.25 μM, and D = 3.125 μM.[91]

Fawn-Hooded Rat

A mild bleeding disorder in fawn-hooded (FH) rats was recognized by Dr. Maier at the University of Michigan.[147] These rats are thought to have arisen at the University of Michigan from breeding German brown rats with white Lashley rats.[145,147] Tschopp and Zucker[147] reported that FH rats had prolonged bleeding times and reduced retention of platelets on glass beads. This hemorrhagic diathesis could not be attributed to a defect in platelet counts nor to a coagulopathy. It was corrected by platelet transfusion. ADP-induced platelet aggregation was normal, but the platelets did not aggregate to collagen and released subnormal amounts of ^{14}C-5-HT. Platelets from FH rats contained reduced amounts of ATP, ADP, and 5-HT. Thrombin induced secretion of ATP and ADP was virtually absent. These and other hemostatic findings are summarized in Table 5. Most studies used an inbred strain of FH rats obtained from

TABLE 5
Hemostatic and Selected Hematological Findings in Fawn Hooded Rats

Normal Parameters	Ref.	Abnormal Parameters	Ref.
Pcv	147	Abnormal Parameters- Increased time	
Platelet count	126 147	or impaired platelet functions tests	
platelet volume	147	Bleeding time	147
PT	126 147	PTT	126
PTT	147		
fibrinogen	126	Platelet retention in	
plasminogen	126	glass-bead columns	147
Factor VII	126	aggregation -ADP	
Factor VIII	126 147	(disaggregation rate)	147
Factor IX	126	-collagen	126 147 159
clot retraction	126 147	-thrombin	147
aggregation-ADP (initial slope)	126 147	-AA	159
-AA, in cholesterol fed rats	159	secretion-reduced ATP, ADP	
		and ^{14}C-5-HT	147
Platelet responses to exposed			
subendothelium *in vivo*		Platelet content of:	
-adhesion rate	146	ATP	126 147
Platelet TXB$_2$ formation	159	ADP	147
		5-HT	55 126 147
Abnormal Parameters- Increased factor			
and shortened coagulation tests		Platelet adhesion to the	
		subendothelium *in vitro*	
Factor IX	126	-adhesion rate	146
Factor X	126	-platelet spreading	146
RVVT-shortened	126	-platelet thrombi formation	146
		Platelet adhesion to the	
		subendotheliun *in vivo*	
		-platelet spreading	146
		-platelet thrombi formation	146

Dr. W. Jean Dodds in Albany, NY, originated from the University of Michigan and were related to the rats studied by Tschopp and Zucker.[147]

In addition to the fawn color coat, FH rats have red eyes, resulting from an absence of melanosomes in the pigmented epithelium.[64] The platelet SPD is related to the fawn color rather than the hood.[121,145] The FH rat is an animal model of SPD, but it is not an animal model of the CHS, since leukocyte granules and melanosomes are not enlarged and natural killer cell activity is not depressed.[140]

The FH rat is very docile.[55,126] Because of the platelet SPD and the unusual behavior, they have been used in several studies of brain 5-HT. A high activity of brain monoamino oxidase for 5-HT[55] and a decreased number of imipramine binding sites in the cerebral cortex have been reported.[95] Total brain levels of tryptophan, kynurenine, 5-HT and 5HIAA were similar to Sprague-Dawley (SD) as was 5-HT turnover,[55] although the brain weight expressed on a body weight basis was lower in FH rats than in SD rats.[55] Other pathological findings of FH rats include an auricular chondritis that was not related to the SPD.[122] A proteinuria that progresses with age was seen in male FH rats.[59] Renal lesions included focal glomerular sclerosis, initially involving the glomerula epithelial cells.[59] There was also spontaneous hypertension.[59]

Miscellaneous Disorders Where Congenital SPD has been Described

Thrombocytopenic Absent Radii (TAR) Syndrome

A platelet SPD in a TAR patient was first described by Day and Holmsen.[26] The content of ADP and 5-HT was comparable to that reported in platelets from HPS patients. This patient was part of a series of SPD patients described by Lages et al.[62] where thrombin and A23187 were used to induce maximal secretion. Releasable adenine nucleotides and calcium were decreased in this patient.

Wiskott-Aldrich Syndrome

Grottum et al.[34] reported on three patients with the Wiskott-Aldrich syndrome. Osmiophilic dense granules were rarely encountered. In one patient, platelet ATP and especially ADP content was reduced. Only negligible amounts of ADP could be released by collagen stimulation, and releasable ATP could not be detected. The survival time of platelets in these patients was only two days.

Porcine SPD

An SPD was recently described in 11 pigs from a colony of pigs with von Willebrand's disease at the Mayo Clinic research facility.[25] These animals were derived from a single boar and three unaffected females, and the defect was recessively transmitted. The SPD pigs had increased bleeding times and severe bleeding disorders. Episodes of epistaxis, dental bleeding and severe hemorrhage following parturition were observed. Collagen-induced aggregation was absent or decreased in most of these animals. Platelet ADP and 5-HT were markedly decreased and platelet ATP was reduced to a lesser degree. This SPD was not as pronounced in these pigs as it is in the HPS, the ep/ep mouse or in animal models of the CHS. The initial uptake of 5-HT was normal while platelet retention of 5-HT after imipramine treatment was reduced. In thin sections of platelets, dense granule numbers were reduced to 26% of normal.

ACQUIRED SPD

Myeloproliferative Disorders

Hemostatic alterations with episodes of bleeding and thrombosis are often present in patients with myeloproliferative disorders. Many of these patients have prolonged bleeding times,[73,94,115] and several have a platelet SPD (Table 6). In most patients, the nucleotide reduction and dense granule deficit[20,31,130] are not as severe as that seen in HPS patients,[73] suggesting that there is a reduction but not an inability to store dense granule constituents. The platelet

TABLE 6
Association of SPD with Different Myeloproliferative Disorders

Syndrome	Number of patients*	Evidence for SPD	Ref.
AMM	3/3	mepacrine granules markedly decreased	132
	3/3	5-HT uptake markedly decreased	132
	3/3	thrombin-induced 5-HT secretion markedly decreased	132
	15/21	collagen-induced aggregation defect	117 94 136* 169
	17/21	epinephrine-induced aggregation defect	169 94 117
	2/2	secretable total adenine nucleotides reduced	136*
	10/10	decreased ADP content	73 94
	4/6	decreased ATP	94
	6/6	decreased 5-HT	94
CGL	31/52	epinephrine-induced aggregation reduced	31 94 115 117 169
(CML)	21/58	collagen-induced aggregation reduced	31 94 115 117 136* 169
	22/27	platelet ADP content is decreased	31 94 115
	12/12	platelet ATP content is reduced	115
	12/12	platelet ATP/ADO ration is increased	115
	5/5	decreased number of dense bodies/ 100 thin sections of platelets	31
	5/5	secretable total adenine nucleotides reduced	136*
	6/10	decreased 5-HT	94

Syndrome	Number of patients*	Evidence for SPD	Ref.
CML	3/3	epinephrine-induced aggregation defect	117
	3/3	collagen-induced aggregation defect	117
CNL	1/1	decreased platelet ADP content	78
	1/1	increased ATP/ADP ratio	78
	1/1	abnormal collagen-induced aggregation	78
	1/1	abnormal epinephrine-induced aggregation	78
PRV	11/25	epinephrine-induced aggregation defect	117 94
	15/25	collagen-induced aggregation defect	94 117 136*
	10/14	decreased platelet ADP content	73 94
	1/1	decreased releasable ADP	73
	2/13	decreased ATP content	94
	10/13	decrease 5-HT	94
MD	4/4	decreased relesable ADP	73
	4/4	decrease total ADP	73
ET	11/28	epinephrine-induced aggregation defect	117 94
	11/31	collagen-induced aggregation defect	117 94 136*
	3/3	secretable adenine nucleotides reduced	136*
	22/23	decreased ADP content	94
	8/23	decreased ATP	94
	16/22	decreased 5-HT	94
Not altered			
ET	3/3	mepacrine containing granules	132
	3/3	5-HT uptake	132
PRV	1/1	mepacrine containing granules	132
	1/1	5-HT uptake	132
CGL	10/10	normal platelet ATP	94

*number of patients exhibiting the abnormality/number of patients tested
ET=essential or primary thrombocythemia
AMM=Agnogenic myeloid metaplasia or myelofibrosis
CGL=chronic granuloctic leukemia or chronic myelogenous leukemia
PRV=polycythemia rubra vera or polycythemia vera
MD=myelodysplastic syndrome
CNL=chronic neutrophilic leukemia
CML=chronic myelomonocytic leukemia

SPD sometimes normalizes spontaneously and this correction is not related to drug therapy or change in the course of the disease.

The SPD may result from abnormal megakaryocytopoiesis. Several MK and platelet ultrastructural abnormalities, including alterations in platelet size, membranes, cytoskeleton and granules, have been described in patients with myeloproliferative disorders.[20,21,40] It has also been suggested that the SPD may result from peripheral blood platelet activation and secretion of dense granule constituents.[94] This scenario appears unlikely since constituents normally found in other platelet granules are not reduced in platelets from these patients.[94]

Neiuwenhuis *et al.*,[98] have reported on 39 patients who had a platelet SPD with myeloproliferative disorders. These cases included five acute (nonlymphocytic) leukemia, three acute lymphocytic leukemia, eight myelodysplastic syndromes, seven chronic myelogenous leukemia, one chronic myelomonocytic leukemia, four polycythemia vera, six essential thrombocythemia, two myelofibrosis, two chronic lymphocytic leukemia, and one Hodgkin's disease patient.

Immunologically Mediated Platelet SPD

Immune mediated platelet injury and destruction may result from antibodies arising spontaneously or associated with other diseases, blood transfusions, pregnancy, or drug treatment. A summary of patients with immunological abnormalities where a platelet SPD may be present is presented in Table 7. Five patients with increased platelet associated IgG (PAIgG) were reported to have a platelet SPD.[153] Beta thromboglobulin, which is a secretable alpha granule constituent, and acid phosphatase, which is a nonsecretable acid hydrolase, were decreased in these patients, while β-N-acetyl glucosaminidase and β-glucuronidase, which are secretable acid hydrolases, were normal.[153] Patients with both systemic lupus erythematosus (SLE) and SPD were described by Pareti *et al.*[113] (n=1), and by Nieuwenhuis *et al.*[98] (n=7). Abnormal second-wave aggregation and reduced total secretable platelet adenine nucleotides were reported in a patient with a complex immunological disorder, including an autoimmune hemolytic anemia.[135] PAIgG was not determined in patients described by Pareti *et al.*[113] and Russell *et al.*[135] Zahavi and Marder[170] reported on a patient with an immunological disorder where there were circulating anti-platelet antibodies and an SPD. In this patient, treatment successfully corrected the thrombocytopenia prior to correcting the prolonged bleeding time.

Hourdille *et al.*[50] reported that 17 out of 22 patients with the chronic idiopathic thrombocytopenia purpura (ITP) had normal or increased levels of platelet ADP. ITP patients also had a slight increase in the mean number of mepacrine containing granules per platelet. Normal platelets contained 6.13+1.05 mepacrine granules/platelet while the ITP patients had 6.89+1.66. In the normal platelet population, the number of mepacrine-containing granules/platelet never exceeded 16. In the ITP group, 10 patients had platelets with more than 16 granules, and some had as many as 22. The platelets with increased numbers of

TABLE 7
Summary of Patients with Imunological Dysfunction
Where SPD Has Been Suggested

Patient	Diagnosis	ATP*	ADP*	5-HT**	Ref.
1	SLE	29	3	45	153
2	SLE	34	10	42	153
3	suspected preleukemia	43	6	40	153
4	chronic ITP	49	9	150	153
5	ND**	41	8	124	153
	normal	44 ± 9	23 ± 4	278	153
6	SLE	31	10	30	113
	normal values	67 ± 7	39 ± 5	300 ± 8	113
7	ND**	37	20	<100	170
	normal values	69	27	300	170

* nmole/10^9 platelets ** nmole/10^9 platelets × 100
** not diagnosed (ND)

mepacrine granules were also very large. The remaining seven patients had high numbers of platelets that lacked mepacrine granules. Platelets from two of these patients had decreased total ADP content.

Alcohol Abuse and SPD

Multiple hemostatic defects are seen with alcohol abuse. The incidence of abnormal hemostasis in patients with alcoholism varies from 24 to over 70%.[19] The platelet abnormalities can occur as a direct effect of alcohol and can be seen in the absence of metabolic or internal organ dysfunction such as liver cirrhosis, splenomegaly or folate deficiency.[19,41] MK maturation is altered with alcohol abuse,[66] and a thrombocytopenia[19] and a thrombopathia[19,41] are often observed. Haut and Cowan[41] reported that thrombin-induced release of ATP and ADP in two alcoholic thrombocytopenic patients was decreased to one-third that of control platelets. This difference was not seen when the patients were not ingesting alcohol nor in another alcoholic patient without thrombocytopenia. In another patient who was ingesting alcohol, total platelet and releasable ATP and ADP were moderately depressed (Table 8).[41]

Cardiopulmonary Bypass (CPB) and SPD

Cardiac surgery utilizing CPB is a surgical procedure with the possibility of serious post-operative bleeding. Bleeding time is consistently and markedly increased during CPB. Multiple platelet lesions have been described, including a platelet SPD. Friedenberg et al.[28] reported that platelets from three CPB patients had a reduced number of dense granules, however, the numbers of dense granules were not quantitated. In an *in vitro* study, platelets were

TABLE 8
Adenine Nucleotides in a Patient with Chronic Alcohol Ingestion**

	ATP	Total ADP	ATP	Releasable ADP	Ref.
Control	5.09*	3.42	0.9	1.86	41
	(0.55)	(0.52)	(0.23)	(0.47)	
Patient	3.95	2.14	0.75	0.95	41

*μmoles/10^{11} , mean (sd)
**from Haut and Cowan[41]

TABLE 9
Adenine Nucleotides in Platelets From
Cardiopulmonary Bypass (CPB) Patients

Ref.	Total ATP	Total ADP	ATP/ADP	Releasable ATP	Releasable ADP
11 (n=64)					
pre	5.5	3.0	1.97	1.38	1.0
	(1.3)	(0.8)	(0.4)	(0.4)	(0.4)
post*	4.9	2.5	2.2	1.0	1.51
	(1.1)	(0.8)	(0.6)	(0.4)	(0.3)
39 (n=4)					
pre	3.39	1.06	1.82		
	(0.98)	(0.8)	(0.33)		
post	2.99	1.82	1.65		
	(0.75)	(0.59)	(0.14)		

Animal studies

	Animal	Total ATP	Total ADP	ATP/ADP	5-HT
39	Baboons				
	Pre				
		1	3.9**	3.5**	1.11
		2	3.5	3.4	1.03
	Post				
		1	3.7	3.3	1.12
		2	3.3	3.1	1.10
6	Dogs				
	Pre				1.45***
	Post				0.395

* sample collected 3.25 + 1.7 hr post surgery
** μg/10^{11}
*** μg/10^{9}

recirculated through a spiral membrane oxygenator similar to that used during CPB.[1] After six hours, the platelets were virtually devoid of granules. Harker *et al.*,[39] on the other hand, found that platelets from two patients after CPB had a normal number of ultrastructurally identifiable dense granules. Biochemical evidence for an SPD with CPB included a modest but statistically significant decrease in platelet ATP and ADP and an increase in ATP/ADP ratio (Table 9).[11] Dogs placed on CPB also have a marked reduction in platelet 5-HT.[6] Harker *et al.*[39] did not find a reduction of platelet nucleotides in four CPB patients nor in two baboons on CPB.

It appears, therefore, that a reduction in agonists stored and released from platelet dense granules is seen in some, but not all, patients undergoing CPB. The reduction is modest and may not, by itself, be responsible for the markedly altered hemostasis.

Miscellaneous Diseases Where Acquired SPD has been Described

Pareti *et al.*[112,113] reported on a series of thrombocytopenic patients associated with increased platelet consumption and an SPD (Table 10). A platelet SPD was observed in three patients with renal transplant rejection (RTR), two patients with hemolytic-uremic syndrome (HUS) and one patient with thrombotic thrombocytopenic purpura (TTP), and his group also reported on two patients with disseminated intravascular coagulation (DIC).[112,113] Four additional cases of DIC and SPD were reported by Nieuwenhuis *et al.*[98] These authors also reported on the SPD in patients with endocarditis (n=1), erythropoietic protoporphyria (n=1), paroxysmal nocturnal hemoglobinuria (n=2), and Waldenstrom's macroglobinemia (n=1).

TABLE 10
Platelet ATP, ADP and 5-HT Content in Patients With Non-immunological Disease and Acquired SPD

Patient	Diagnosis	ATP*	ADP	5-HT**	ref.
1.	RTR	53	17	30	113
2.	RTR	64	15	80	113
3.	RTR	98	30	70	113
4.	HUS	—	—	60	113
5.	HUS	87	25	40	113
6.	TTP	—	—	20	113
7.	DIC	47	16	70	113
8.	DIC	52	19	70	112
	CONTROLS	67	39	370	113
		69	39	300	112

* nmole/10^9 platelets ** nmole/10^9 platelets × 100

TREATMENT

Treatment to control hemorrhage may become necessary in some SPD patients. Treatments have focused on either enhancing overall hemostatic competency or providing normal platelets. Treatments to enhance hemostatic competency include cyroprecipitate[33] and the vasopressin analog 1-deamino-8-D-arginine vasopressin (DDAVP).[57,76,99]

Cryoprecipitate is a concentrate of Factor VIII and von Willebrand factor (vWf) that is prepared by a controlled thaw of fresh frozen plasma. The first SPD patient treated with cryoprecipitate was initially diagnosed as having a vWf deficiency, and the bleeding time shortened after treatment.[33] Gerritsen *et al.*[33] subsequently treated eight additional SPD patients with cryoprecipitate, five with HPS and three with SPD without albinism. Following treatment, the bleeding times shortened but did not normalize in every patient. Shortening of the bleeding times occurred without correction of the SPD.

DDAVP induces a rapid increase in plasma vWf and Factor VIII.[75] The vWf circulates as a series of multimers, varying from 0.5 to 20×10^6 Da. The largest multimers are the most physiologically active and increase preferentially following DDAVP administration. DDAVP treatment has been reported to shorten the bleeding time for two to four hours in some,[57,99,163] but not all[57,76,99] SPD patients. Mannucci *et al.*[76] reported on administering 0.3 µg/kg of DDAVP to seven SPD patients. While the median bleeding time did not significantly change, bleeding time did shorten in three patients. Nieuwenhuis and Sixma[99] reported on treating 19 patients with congenital SPD and 4 patients with acquired SPD with 0.4 µg/kg of DDAVP. Bleeding time shortened in 18 of these patients and was unchanged in 5 others. In three HPS patients, treatment with DDAVP (0.3 µg/kg) shortened the bleeding time in two and was inconclusive in the third patient.[163] The reason why some SPD patients respond to DDAVP while others do not is not readily apparent.

DDAVP has disadvantages, including tachyphylaxis[75] and elevating blood plasminogen activators.[74] Cryoprecipitate[33] and DDAVP[99] have been used prophylactically prior to surgery, and excessive bleeding was not reported in these cases.

Correction of primary hemostasis in SPD patients can be achieved with platelet transfusions[22,147] or bone marrow transplantation.[103] Novak *et al.*[103] were able to markedly shorten the prolonged bleeding time in pale and beige mice by transplanting bone marrow from normal congenic mice. Platelet transfusion was stated to be effective in treating human SPD[33] and to be the treatment of choice when bleeding becomes life threatening.[99] Platelet transfusion has been shown to correct the bleeding time in fawn-hooded rats.[147] Transfusing CHS cats with normal feline platelets shortens the buccal mucosal bleeding time from over 9 minutes to less than 2.5 minutes, with a donor platelet half life of 3.5 days.[22] The bleeding time was normalized when the donor platelet count in the CHS cats increased to 50,000 platelets/µl whole blood. Many of the drawbacks previously associated with platelet transfusion have been reduced by improving diagnostic tests for transmissible diseases and by decreasing leukocyte contamination to postpone isoimmunization.

SUMMARY

Patients with platelet SPD have a mild to moderate bleeding diathesis. However, severe life-threatening or even fatal bleeding episodes may occur. The bleeding problem is exacerbated when the conversion of AA to TxA_2 has been inhibited. Patients with SPD can have platelet aggregation abnormalities such as impaired aggregation at low concentrations of collagen, platelet disaggregation and the absence of a biphasic aggregation wave. However, biphasic platelet aggregation is seen in a substantial number of patients. Normal aggregation responses have been attributed to an increased sensitivity to platelet ADP or to TxA_2 formation. The AA pathway appears to be altered in some SPD patients and animal models but not in others.

SPD can be congenital or acquired. With congenital SPD, the deficiency can be complete or partial. In some SPD patients only dense granule deficiency may be observed, while in others, platelet alpha granules may also be reduced. There are several animal models of platelet SPD. Many human patients and animal models have a melanosome and lysosome abnormality as well as an SPD, indicating that these three organelles share similar biosynthetic processes. An SPD could be due to a physiological defect where the granule is present but cannot store adenine nucleotides and 5-HT. On the other hand, the SPD could be anatomical and the dense granule or its precursor is not within the platelet.

CHS cattle are one of the most extensively studied animal models and, based on ultrastructural, biochemical and functional studies, there is no evidence that the dense granule or its precursor is present in CHS platelets. The platelet dense granule deficiency in CHS cattle cannot be attributed to enhanced granule fusion or abnormal processing within the MK. Since the dense granule precursors cannot be identified in immature, maturing or mature MK, the most plausible explanation for the SPD is that dense granules are not formed.

Acquired SPDs are associated with myeloproliferative disorders, conditions where megakaryocytoporesis is altered, and in conditions where increased platelet destruction or consumption occurs. Acquired SPD is usually a partial rather than a total deficiency. It is not known whether the partial SPD is due to some platelets being normal while others are severely deficient or if all platelets are affected.

Treatment for SPD includes platelet transfusion, DDAVP administration, cryoprecipitate and bone marrow transplantation.

List of Abbreviations

AA = Arachidonic acid
CHS = Chediak-Higashi syndrome
CPB = Cardiopulmonary bypass
DDAVP = 1-deamino-8-D-arginine vasopressin
DIC = Disseminated intravascular coagulation
GFP = Gel filtered platelets
HNF = Heparin-neutralizing factor
HPS = Hermansky-Pudlak syndrome
HUS = Hemolytic-uremic syndrome
ITP = Idiopathic thrombocytopenia purpura
MBS = Membrane bound structures
MDA = Malonaldehyde
MK = Megakaryocyte
PAF = Platelet activating factor
RTR = Renal transplant rejection
SDS = Sodium dodecyl sulfate
SPD = Storage pool deficiency
SLE = Systemic lupus erythematosus
TAR = Thrombocytopenic absent radii
TTP = Thrombotic thrombocytopenic purpura
vWf = von Willebrand factor

REFERENCES

1. Addonizio, V.P., Marcarak, E.J., Niewiarowski, S., Colman, R.W., and Edmunds, L.H., Preservation of human platelets with prostaglandin E1 during *in vitro* stimulation of cardiopulmonary bypass, *Circ. Res.*, 44, 350-357, 1979.

2. Akkerman, J.W., Nieuwenhuis, H.K., Mommersteeg-Leautaud, M.E., Gorter, G., and Sixma, J.J., ATP-ADP compartmentation in storage pool deficient platelets: Correlation between granule-bound ADP and the bleeding time, *Br. J. Haematol.*, 55, 135-143, 1983.

3. Allison, A.C., and Young, M.R., Uptake of dyes and drugs by living cells in culture. *Life Sci.*, 3, 1407-1414, 1964.

4. Apitz-Castro, R., Cruz, M.R., Ledezman, E., Merino, F., Ramerez-Duque, C., Dangelmeier, C., and Holmsen, H., The storage pool deficiency in platelets from humans with the Chediak-Higashi syndrome: A study of six patients, *Br. J. Haematol.*, 59, 471-483, 1985.

5. Ayers, J.R., Leipold, H.W., and Padgett, G.A., Lesions in Brangus cattle with Chediak-Higashi syndrome, *Vet. Pathol.*, 25, 432-436, 1988.

6. Aznavoorian, S.A., Utsunomiya, Y., Krausz, M.M., Cohn, L.H., Shepro, D., and Hechtman, H.B., Prostacyclin inhibits 5-hydroxytryptamine release but stimulates thromboxane synthesis during cardiopulmonary bypass, *Prostaglandins*, 25, 557-570, 1983.

7. Bartholini, G.A., and Pletscher, A., Formation of 5-hydroxytryptophol from endogenous 5-hydroxytryptamine by isolated blood platelets, *Nature*, 203, 1281-1283, 1964.

8. Bell, T.G., Camacho, Z., Meyers, K.M., and Padgett, G.A., Platelet storage disease in the Chediak-Higashi syndrome, *Fed. Proc.*, 34, 861, 1975.

9. **Bell, T.G., Meyers, K.M., Prieur, D.J., Fauci, A.S., Wolff, S.M., and Padgett, G.A.,** Decreased nucleotide and serotonin storage associated with defective function in Chediak-Higashi syndrome cattle and human platelets, *Blood,* 48, 175-183, 1976.
10. **Bequez-Cesar, A.,** Neutropenia cronica maligna familiar con granulaciones atipics de lost leucocitos, *Sociedad Cubana de Pediatr Boletin,* 15, 900-922, 1943.
11. **Beurling-Harbury, D., and Galvan, C.A.,** Acquired decrease in platelet secretory ADP associated with increased postoperative bleeding in post-cardiopulmonary bypass patients and in patients with severe valvular disease, *Blood,* 54, 13-23, 1974.
12. **Blume, R.S., and Wolff, S.M.,** the Chediak-Higashi syndrome: Studies in four patients and a review of the literature, *Medicine,* 51, 247-280, 1972.
13. **Boxer, G.J., Holmsen, H., Robkin, L., Bank, N.Y., Boxer, L.A., and Baehner, R.L.,** Abnormal platelet functions in the Chediak-Higashi syndrome, *Br. J. Haematol.,* 35, 521-533, 1977.
14. **Boxer, L.A., Albertini, D.E., Baehner, R.L., and Oliver, J.M.,** Correction of leukocyte functions in the Chediak-Higashi syndrome correctable by ascorbate, *N. Eng. J. Med.,* 43, 207-213, 1976.
15. **Buchanan, G.R., and Handin, R.I.,** Platelet function in the Chediak-Higashi syndrome, *Blood,* 47, 941-948, 1976.
16. **Chediak, M.M.,** Nouvelle anomalie leucocytaire de caracttere constituitionnel et familial, *Rev. Hematol.,* 7, 362-367, 1952.
17. **Colgan, S.P., Hull-Thrall, M.A., and Gasper, P.W.,** Platelet aggregation and ATP secretion in whole blood of normal cats and cats homozygous and heterozygous for Chediak-Higashi syndrome, *Blood Cells,* 15, 585-595, 1989.
18. **Costa, J.L., Fauci, A.S., and Wolf, S.M.,** A platelet abnormality in the Chediak-Higashi syndrome, *Blood,* 48, 517-520, 1976.
19. **Cowan, D.H.,** Effect of alcoholism on hemostasis, *Semin. Hematol.,* 17, 137-147, 1980.
20. **Cowan, D.H., and Graham, R.C.,** Structural-functional relationships in platelets in acute leukemia and related disorders, *Ser. Haematol.,* 8, 68-100, 1975.
21. **Cowan, D.H., Graham, R.C., and Baunach, D.,** The platelet defect in leukemia: Platelet ultrastructure, adenine nucleotide metabolism, and the release reaction, *J. Clin. Invest.,* 56, 188-200, 1975.
22. **Cowles, B.E., Meyers, K.M., Wardrop, K.J., Menard, M., and Sylvester, D.,** Transfusion of normal feline platelets corrects the bleeding time in cats with the Chediak-Higashi syndrome, *Thromb. Haemost,* in press, 1991.
23. **Da Prada, M., Picotti, G.B., Kettler, R., and Launay, J.M.,** Amine storage organelles in platelets,. in *Platelets in Biology and Medicine,* 2nd Ed., Gordon, J.L., Ed., North-Holland, New York, 1981, pp. 277-288.
24. **Daniel, J.L., Molish, I.R., and Holmsen, H.,** Radioactive labeling of the adenine nucleotide pool of cells as a means to distinguish among intracellular compartments. Studies on human platelets, *Biochem. Biophys. Acta,* 632, 444-453, 1980.
25. **Daniels, T.M., Fass, D.N., White, J.G., and Bowie, E.J.W.,** Platelet storage pool disease deficiency in pigs, *Blood,* 67, 1043-1047, 1986.
26. **Day, H.J., and Holmsen, H.,** Platelet adenine nucleotide "storage pool deficiency" in thrombocytopenic absent radii syndrome, *JAMA,* 221, 1053-1054, 1972.
27. **Fagerland, J.A., Hagemoser, W.A., and Ireland, W.P.,** Ultrastructure of resting and serology of leukocytes and platelets of normal foxes and a fox with a Chediak-Higashi-like syndrome, *Vet. Pathol.,* 24, 164-169, 1987.
28. **Friedenberg, W.R., Myers, W.O., Plotka, E.D., Beathard, J.N., Kummer, D.J., Gatlin, P.F., Stoiber, D.L., Ray, J.F., and Sautter, R.D.,** Platelet dysfunction associated with cardiopulmonary bypass, *Ann. Thorac. Surg.,* 25, 298-305, 1978.
29. **Fukami, M.H., Bauer, J.S., Stewart, G.J., and Salganicoff, L.,** An improved method for the isolation of dense storage granules from human platelets, *J. Cell Biol.,* 77, 389-399, 1978.

30. **Gerrard, J.M., Lint, D., Sims, P.J., Wiedmer, T., Fugate, R.D., McMillan, E., Robertson, C., and Israels, S.J.**, Identification of a platelet dense granule membrane protein that is deficient in a patient with the Hermansky-Pudlak syndrome, *Blood*, 77, 101-112, 1991.

31. **Gerrard, J.M., Stoddard, S.F., Shapiro, R.S., Coccia, P.F., Ramsay, N.K.C., Nesbit, M.E., Rao, G.H.R., Krivit, W., and White, J.G.**, Platelet storage pool deficiency and prostaglandin synthesis in chronic granulocytic leukaemia, *Br. J. Haematol.*, 40, 597-607, 1978.

32. **Gerritsen, S.M., Akkerman, J.W., Nijeijer, B., Sixma, J.J., Witkop, C.J., and White, J.**, The Hermansky-Pudlak syndrome. Evidence for a lowered 5-hydroxytryptamine content in platelets of heterozygotes, *Scand. J. Haematol.*, 18, 249-256, 1977.

33. **Gerritsen, S.W., Akkerman, J.W., and Sixma, J.J.**, Correction of the bleeding time in patients with storage pool deficiency by infusion of cryoprecipitate, *Br. J. Haematol.*, 40, 153-160, 1978.

34. **Grottum, K.A., Hovig, T., Holmsen, H., Foss Abrahamsen, A., Jeremic, M., and Seip, M.**, Wiskott-Aldrich syndrome: Qualitative platelet defects and short platelet survival, *Br. J. Haematol.*, 17, 373-388, 1969.

35. **Haak, R.A., Ingraham, L.M., Baechner, R.L., and Boxer, L.A.**, Membrane fluidity in human and Chediak-Higashi leukocytes, *J. Clin. Invest.*, 64, 138-144, 1979.

36. **Hardisty, R.M., and Hutton, R.A.**, Bleeding tendency associated with a "new" abnormality of platelet behavior, *Lancet*, 1, 983-985, 1967.

37. **Hardisty, R.M., Mills, D.C.B., and Ketsaard, K.**, The platelet defect associate with albinism, *Br. J. Haematol.*, 23, 679-692, 1972.

38. **Hardisty, R.M., and Stacy, R.S.**, 5-hydroxytryptamine in normal human platelets, *J. Physiol. (London)*, 130, 711-720, 1955.

39. **Harker, L.A., Malpass, T.W., Branson, H.E., Hessel, E.A., and Slichter, S.J.**, Mechanism of abnormal bleeding in patients undergoing cardiopulmonary bypass: Acquired transient platelet dysfunction associated with selective alpha-granule release, *Blood*, 56, 824-834, 1980.

40. **Hattori, A., Koike, K., and Ito, S.**, Static and functional morphology of pathological platelets in primary myeofibrosis and myeloproliferative syndrome, *Ser. Haematol.*, 8, 126-150, 1975.

41. **Haut, M.J., and Cowan, D.H.**, The effects of ethanol on hemostatic properties of human blood platelets, *Am. J. Med.*, 56, 22-33, 1974.

42. **Hermansky, F., and Pudlak, P.**, Albinism associated with hemorrhagic diathesis and unusual pigmented reticular cells in the bone marrow: Report of two cases with histochemical studies, *Blood*, 14, 162-169, 1959.

43. **Higashi, O.**, Congenital giantism of peroxidase granules, *Tohoku J. Exp. Med.*, 59, 315-332, 1954.

44. **Holland, J.M.**, Studies of membranes in animals with the Chediak-Higashi syndrome, *Doctoral Dissertation*, College of Veterinary Medicine, Washington State University, Pullman, WA., 1970.

45. **Holland, J.M.**, Serotonin deficiency and prolonged bleeding time in beige mice, *Proc. Soc. Exp. Biol. Med.*, 151, 32-39, 1976.

46. **Holmsen, H., Setkowsky, C.A., Lages, B., Day, H.J., Weiss, H.J., and Scrutton, M.C.**, Content and thrombin-induced release of acid hydrolases in gel-filtered platelets from patients with storage pool disease, *Blood*, 46, 131-142, 1975.

47. **Holmsen, H., Storm, E., and Day, H.J.**, Determination of ATP and ADP in blood platelets, *Analyt. Biochem.*, 46, 489-501, 1972.

48. **Holmsen, H., and Weiss, H.J.**, Hereditary defect in the platelet release reaction caused by a deficiency in the storage pool of platelet adenine nucleotides, *Br. J. Haematol.*, 19, 643-649, 1970.

49. **Holmsen, H., and Weiss, H.J.**, Further evidence for a deficient storage pool of adenine nucleotides in platelets from some patients with thrombopathia-"Storage pool disease," *Blood*, 39, 197-209, 1972.

50. **Hourdille, P., Bernard, P., Reiffers, J., Broustet, A., and Boisseau, M.R.**, Platelet dense bodies loaded with mepacrine study in chronic idiopathic thrombocytopenia purpura (ITP), *Thromb. Haemost.*, 43, 208-210, 1980.

51. **Ingerman, C.M., Smith, J.B., Shapiro, S., Sedar, A., and Silver, M.J.**, Hereditary abnormality of platelet aggregation attributable to nucleotide storage pool deficiency, *Blood*, 52, 332-344, 1978.

52. **Israels, S.J., McNicol, A., Robertson, C., and Gerrard, J.M.**, Platelet storage pool deficiency: Diagnosis in patients with prolonged bleeding times and normal platelet aggregation, *Br. J. Haematol.*, 75, 118-121, 1990.

53. **Ito, M., Sato, A., Tanabe, F., Ishida, E., Takami, Y., and Shigeta, S.**, The thiol proteinase inhibitors improve the abnormal rapid down-regulation of protein kinase C and the impaired natural killer cell activity in (Chediak-Higashi syndrome) beige mouse, *Biochem. Biophys. Res. Comm.*, 160, 433-440, 1989.

54. **Jamieson, G.A., Okumura, T., Fishback, B., Johnson, M.M., Egan, J.J., and Weiss, H.J.**, Platelet membrane glycoproteins in thrombasthemia, Bernard-Soulier syndrome, and storage pool disease, *J. Lab. Clin. Med.*, 93, 652-660, 1979.

55. **Joseph, M.H.**, Brain tryptophan metabolism on the 5-hydroxytryptamine and kynurenine pathways in a strain of rats with a deficiency in platelet 5-HT, *Br. J. Pharmacol.*, 63, 529-533, 1978.

56. **Kimbal, H.R., Ford, G.H., and Wolff, S.M.**, Lysosomal enzymes in normal and Chediak-Higashi blood leukocytes, *J. Lab. Clin. Med.*, 86, 616-630, 1975.

57. **Kobrinsky, N.L., Israels, E.D., Gerrard, J.M., Cheang, M.S., Watson, C.M., Bishop, A.J., and Schroeder, M.L.**, Shortening of bleeding time by 1-deamino-8-d-arginine vasopressin in various bleeding disorders, *Lancet*, 1, 1145-1148, 1984.

58. **Kramer, J.W., Davis, W.C., and Prieur, D.J.**, The Chediak-Higashi syndrome of cats, *Lab Invest.*, 36, 554-562, 1977.

59. **Kreisberg, J.I., and Karnovsky, M.J.**, Focal glomerular sclerosis in the fawn-hooded rat, *Am. J. Pathol.*, 92, 637-652, 1978.

60. **Lages, B., Dangelmaier, C.A., Holmsen, H., and Weiss, H.J.**, Specific correction of impaired acid hydrolase secretion in storage pool-deficient platelets by adenosine diphosphate, *J. Clin. Invest.*, 81, 1865-1872, 1988.

61. **Lages, B., Holmsen, H., Scrutton, M.C., Day, H.J., and Weiss, H.J.**, Metal ion content of gel filtered platelets from patients with storage pool disease, *Blood*, 469, 119-130, 1975.

62. **Lages, B., Holmsen, H., Weiss, H.J., and Dangelmaier, C.**, Thrombin and ionophore A23187-induced dense granule secretion in storage pool deficient platelets: Evidence for impaired nucleotide storage as the primary dense granule defect, *Blood*, 61, 154-162, 1983.

63. **Lages, B., and Weiss, H.J.**, Biphasic aggregation responses to ADP and epinephrine in some storage pool deficient platelets: Relationship to the role of endogenous ADP in platelet aggregation and secretion, *Thromb. Haemost.*, 43, 147-153, 1980.

64. **Lavail, M.M.**, Fawn-hooded rats, the fawn mutation and interaction of pink-eyed and red-eyed dilution genes, *J. Hered.*, 72, 286-287, 1981.

65. **Leader, R.W., Padgett, G.A., and Gorham, J.R.**, Studies of abnormal leukocyte bodies in mink, *Blood*, 22, 477-484, 1963.

66. **Levine, R.F., Spivak, J.L., Meagher, R.C., and Sieber, F.**, Effect of ethanol on thrombocytopoiesis, *Br. J. Haematol.*, 62, 345-354, 1986.

67. **Lorez, H.P., and Da Prada, M.**, Fluorescence microscopical study of 5-hydroxytryptamine storage organelles in mepacrine-incubated blood platelets of beige mice, *Experimentia*, 34, 663-664, 1978.

68. **Lorez, H.P., Richards, J.G., Da Prada, M., Picotti, G.B., Pareti, F.I., Capitanio, A., and Mannucci, P.M.**, Storage pool disease: Comparative fluorescence microscopical, cytochemical and biochemical studies an amine-storing organelles of human blood platelets, *Br. J. Haematol.*, 43, 297-305, 1979.

69. **Lutzner, M.A., Lowrie, C.T., and Jordan, H.W.,** Giant granules in leukocytes of the beige mouse, *J. Hered.*, 58, 299-300, 1967.

70. **Lutzner, M.A., Tierney, J.H., and Benditt, E.P.,** Giant granules and widespread cytoplasmic inclusions in a genetic syndrome and Aleutian minks: An electron microscopic study, *Lab. Invest.*, 14, 2063-2079, 1965.

71. **Lynch, S.R., and Cook, J.D.,** Interaction of vitamin C and iron, *Ann. NY Acad. Sci., USA*, 355, 32-44, 1980.

72. **Malmsten, C., Kindahl, H., Samuelsson, B., Levy-Toledano, S., Tobelem, G., and Caen, J.P.,** Thromboxane synthesis and the platelet release reaction in Bernard-Soulier syndrome, thrombasthenia Glanzmann and Hermansky-Pudlak syndrome, *Br. J. Haematol.*, 35, 511-520, 1977.

73. **Malpass, T.W., Savage, B., Hanson, S.R., Slichter, S.J., and Harker, L.A.,** Correlation between prolonged bleeding time and depletion of platelet dense granule ADP in patients with myelodysplastic and myeloproliferative disorders, *J. Lab. Clin. Med.*, 103, 894-904, 1984.

74. **Mannucci, P.M., Aberg, M., Milsson, I.M., and Robertson, B.,** Mechanism of plasminogen activator and factor VIII increase after vasoactive drugs, *Br. J. Haematol.*, 30, 81-93, 1975.

75. **Mannucci, P.M., Canciani, M.T., Rota, L., and Donovan, B.S.,** Response of factor VIII/ von Willebrand factor to DDAVP in healthy subjects and patients with haemophilia A and von Willebrand's disease, *Br. J. Haematol.*, 47, 283-293, 1981.

76. **Mannucci, P.M., Vicente, V., Vianello, L., Cattaneo, M., Alberca, I., Coccato, M.P., Faioni, E., and Mari, D.,** Controlled trial of desmopressin in liver cirrhosis and other conditions associated with a prolonged bleeding time, *Blood*, 67, 1148-1153, 1986.

77. **Maurer, H.M., Wolff, J.A., Buckingham, S., and Spielvogel, A.R.,** Impotent platelets in albinos with prolonged bleeding times, *Blood*, 34, 204-215, 1972.

78. **Mehrotra, P.K., Winfield, D.A., and Fergusson, L.H.,** Cellular abnormalities and reduced colony-forming cells in chronic neutrophilic leukaemia, *Acta Haematol. (Basel)*, 73, 47-50, 1985.

79. **Meisler, M., Levy, J., Sansone, F., and Gordon, M.,** Morphologic and biochemical abnormalities of kidney lysosomes in mice with an inherited albinism, *Am. J. Pathol.*, 101, 581-594, 1980.

80. **Ménard, M., and Meyers, K.M.,** Storage pool deficiency in cattle with the Chediak-Higashi syndrome results from an absence of dense granule precursors in their megakaryocytes, *Blood*, 72, 1726-1734, 1989.

81. **Ménard, M., Meyers, K.M., and Prieur, D.J.,** Primary and secondary lysosomes in megakaryocytes and platelets from cattle with the Chediak-Higashi syndrome, *Thromb. Haemost.*, 64, 156-160, 1990.

82. **Meyers, K.M.,** Species differences, in *Platelet Responses and Metabolism*, Vol. 1., Holmsen, H., Ed., CRC Press, Inc., Boca Raton, 1986, pp. 209-234.

83. **Meyers, K.M., and Chen, M.,** Brain serotonin concentration and crude synaptosomal uptake in mice with the Chediak-Higashi syndrome, *Neurology*, 26, 1169-1172, 1976.

84. **Meyers, K.M., Costa, J.L., and Magnuson, J.,** Compartmentation of 4,6-difluoro-5HT studies by nuclear magnetic resonance in normal and CHS bovine platelets, *Thromb. Res.*, 58, 265-272, 1990.

85. **Meyers, K.M., Holmsen, H., and Seachord, C.L.,** Comparative study of platelet dense granule constituents, *Am. J. Physiol.*, 243, R454-R461, 1982.

86. **Meyers, K.M., Holmsen, H., Seachord, C.L., Hopkins, G.E., Borchard, R.E., and Padgett, G.A.,** Storage pool deficiency in platelets from Chediak-Higashi cattle, *Am. J. Physiol.*, 237, R239-R248, 1979.

87. **Meyers, K.M., Holmsen, H., Seachord, C.L., Hopkins, G.E., and Gorham, J.,** Characterization of platelets from normal mink and mink with the Chediak-Higashi syndrome, *Am. J. Hematol.*, 7, 137-146, 1979.

88. **Meyers, K.M., Hopkins, G., Holmsen, H., Benson, K., and Prieur, D.J.,** Ultrastructure of resting and activated storage pool deficient platelets from animals with the Chediak-Higashi syndrome, *Am. J. Pathol.*, 106, 364-377, 1982.

89. **Meyers, K.M., and Seachord, C.L.,** Identification of dense granule specific membrane proteins in bovine platelets that are absent in the Chediak-Higashi syndrome, *Thromb. Haemost.*, 64, 319-325, 1990.

90. **Meyers, K.M., Seachord, C.L., Benson, K., Fukami, M., and Holmsen, H.,** Serotonin accumulation in granules of storage pool-deficient platelets from Chediak-Higashi cattle, *Am. J. Physiol.*, 245, H150-H158, 1983.

91. **Meyers, K.M., Seachord, C.L., Holmsen, H., and Prieur, D.J.,** Evaluation of the platelet storage pool deficiency in the feline counterpart of the Chediak-Higashi syndrome, *Am. J. Hematol.*, 11, 241-253, 1981.

92. **Meyers, K.M., Seachord, C.L., Holmsen, H., Smith, B.J., and Prieur, D.J.,** A dominant role for thromboxane formation in secondary aggregation of platelets, *Nature*, 282, 331-333, 1979.

93. **Meyers, K.M., Stevens, D., and Padgett, G.A.,** A platelet serotonin anomaly in the Chediak-Higashi syndrome, *Res. Commun. Chem. Pathol. Pharmacol.*, 2, 375-380, 1974.

94. **Mohri, H.,** Acquired von Willebrands disease and storage pool disease in chromin myelocytic leukemia, *Am. J. Hematol.*, 22, 391-401, 1986.

95. **Murray, T.F., DeBarrows, B.R., Prieur, D.J., and Meyers, K.M.,** [^{3}H]-imipramine binding sites in fawn-hooded rats, *Neuropharmacology*, 22, 781-784, 1983.

96. **Nes, N., Lium, B., and Sjaastad, O.,** A Chediak-Higashi-like syndrome in Artic blue foxes, *Finsk Veterinaertidskrift*, 89, 313, 1983.

97. **Nes, N., Lium, B., Sjaastad, O., Blom, A., and Lohi, O.,** A Norwegian pearl fox (omberg pearl) with Chediak-Higashi syndrome and its relationship to other pearl mutations, *Scientifur*, 9, 197-199, 1985.

98. **Nieuwenhuis, H.K., Akkerman, J.-W.N., and Sixma, J.J.,** Patients with prolonged bleeding time and normal aggregation test may have storage pool deficiency: Studies on one hundred six patients, *Blood*, 70, 620-623, 1987.

99. **Nieuwenhuis, H.K., and Sixma, J.J.,** 1-Deamino-8-D-arginine vasopressin (Desmopressin) shortens the bleeding time in storage pool deficiency, *Ann. Intern. Med.*, 108, 65-67, 1988.

100. **Nishimura, A.M., Inoue, M., Nishikawa, T., Miyamoto, M., Kobayashi, T., and Kitamura, Y.,** Beige rat: A new animal model of Chediak-Higashi syndrome, *Blood*, 74, 270-273, 1989.

101. **Novak, E.K., Hui, S.W., and Swank, R.T.,** The mouse pale ear pigment mutant as a possible animal model for human platelet storage pool deficiency, *Blood*, 57, 38-43, 1981.

102. **Novak, E.K., Hui, S.W., and Swank, R.T.,** Platelet storage pool deficiency in mouse pigment mutations associated with seven distinct genetic loci, *Blood*, 63, 536-544, 1984.

103. **Novak, E.K., McGarry, M.P., and Swank, R.T.,** Correction of symptoms of platelet storage pool deficiency in animal models for Chediak-Higashi syndrome and Hermansky-Pudlak syndrome, *Blood*, 66, 1196-1201, 1985.

104. **Novak, E.K., and Swank, R.T.,** Lysosomal dysfunctions associated with mutations at mouse pigment genes, *Genetics*, 92, 189, 1979.

105. **Novak, E.W., Sweet, H.O., Prochazka, M., Parentis, M., Soble, R., Reddington, M., Chairo, A., and Swank, R.T.,** Coco: A new mouse model for platelet storage pool deficiency, *Br. J. Haematol.*, 69, 371-378, 1988.

106. **Oliver, C., and Essner, E.,** Formation of anomalous lysosomes in monocytes, neutrophils, and eosinophils from bone marrow of mice with the Chediak-Higashi syndrome, *Lab. Invest.*, 32, 17-27, 1975.

107. **Oliver, C., Essner, E., Zimring, A., and Haimes, H.,** Age-related accumulation of ceroid-like pigment in mice with Chediak-Higashi syndrome, *Am. J. Pathol.*, 84, 225-238, 1976.

108. **Oliver, J.M., and Zurier, R.B.,** Correction of characteristics abnormalities of microtubule function and granule morphology in Chediak-Higashi syndrome with cholinergic agonists: Studies *in vitro* in man and *in vivo* in the beige mouse, *J. Clin. Invest.*, 57, 1239-1247, 1976.

109. **Padgett, G.A.,** The Chediak-Higashi syndrome, *Adv. Vet. Sci.,* 12, 239-284, 1968.

110. **Padgett, G.A., Leader, R.W., and Gorham, J.R.,** Familial occurrence of the Chediak-Higashi syndrome in mink and cattle, *Genetics,* 49, 505-512, 1964.

111. **Page, A.R., Berendes, H., Warner, J., and Good, R.A.,** The Chediak-Higashi syndrome, *Blood,* 20, 330-334, 1962.

112. **Pareti, F.I., Capitanio, A., and Mannucci, P.M.,** Acquired storage pool deficiency in platelets during disseminated intravascular coagulation, *Blood,* 48, 511-515, 1976.

113. **Pareti, F.I., Capitanio, A., Mannucci, L., Ponticelli, C., and Mannucci, P.M.,** Acquired dysfunction due to the circulation of "exhausted" platelets, *Am. J. Med.,* 69, 235-240, 1980.

114. **Pareti, F.I., Day, H.J., and Mills, D.C.B.,** Nucleotide and serotonin metabolism in platelets with defective secondary aggregation, *Blood,* 44, 789-800, 1974.

115. **Pareti, F.I., Gugliotta, L., Mannucci, L., Guarini, A., and Mannucci, P.M.,** Biochemical and metabolic aspects of platelet dysfunction in chronic myeloproliferative disorders, *Thromb. Haemost.,* 47, 84-89, 1982.

116. **Parmley, R.T., Poon, M.C., Crist, W.M., and Malhuh, A.,** Giant platelet granules in a child with the Chediak-Higashi syndrome, *Am. J. Hematol.,* 6, 51-60, 1979.

117. **Phadke, K., Dean, S., and Pitney, W.R.,** Platelet dysfunction in myeloproliferative syndromes, *Am. J. Hematol.,* 10, 57-64, 1981.

118. **Phillips, L.L., Kaplan, H.S., Padgett, G.A., and Gorham, J.R.,** Comparative studies on the Chediak-Higashi syndrome. Coagulation and fibrinolytic mechanisms of mink and cattle, *Am. J. Vet. Clin. Path.,* 1, 1-6, 1967.

119. **Piccini, A.E., Jahreis, G.P., Novak, E.K., and Swank, R.T.,** Intracellular distribution of lysosomal enzymes in the mouse pigment mutants pale ear and pallid, *Mol. Cell Biochem.,* 31, 89-95, 1980.

120. **Prieur, D.J., Holland, J.M., Bell, T.G., and Young, D.M.,** Ultrastructure and morphometric studies of platelets with the Chediak-Higashi syndrome, *Lab. Invest.,* 35, 197-204, 1976.

121. **Prieur, D.J., and Meyers, K.M.,** Genetics of the fawn-hooded rat strain. The coat color dilution and platelet storage pool deficiency are pleiotropic effects of the autosomal recessive red-eyed dilution gene, *J. Hered.,* 75, 349-352, 1984.

122. **Prieur, D.J., Young, D.M., and Counts, D.F.,** Auricular chondritis in fawn-hooded rats. A spontaneous disorder resembling that induced by immunization with type II collagen, *Am. J. Pathol.,* 116, 69-76, 1984.

123. **Pyrzwansky, K.B., Schliwa, M., and Boxer, L.A.,** Microtubule organization of unstimulated and stimulated adherent human neutrophils in Chediak-Higashi syndrome, *Blood,* 66, 1398-1403, 1985.

124. **Rao, A.K., Willis, J., Hassell, B., Dangelmaier, C., Holmsen, H., and Smith, J.B.,** Platelet-activating factor is a weak platelet agonist: Evidence from normal human platelets and platelets with congenital secretion defects, *Am. J. Hematol.,* 17, 153-165, 1984.

125. **Rao, G.H., Peller, J.D., and White, J.G.,** Rapid separation of platelet nucleotides by reverse-phase, isocritic, high-performance liquid chromatography with a compressed column, *J. Chromatography,* 226, 466-470, 1981.

126. **Raymond, S.L., and Dodds, W.J.,** Characterization of the fawn-hooded defect in storage pool deficient platelets from fawn-hooded rats, *Thromb. Diath. Haemorrh.,* 33, 361-369, 1975.

127. **Reddington, M., Novak, E.K., Hurley, E., Medda, C., McGarry, M.P., and Swank, R.T.,** Immature dense granules in platelets from mice with platelet storage pool disease, *Blood,* 69, 1300-1306, 1987.

128. **Rendu, F., Brenton-Gorius, J., Lebert, M., Klebanoff, C., Buriot, D., Griscelli, C., Levy-Toledano, S., and Caen, J.P.,** Evidence that abnormal platelet functions in human Chediak-Higashi syndrome are the results of a lack of dense granules, *Am. J. Pathol.,* 111, 309-314, 1983.

129. **Rendu, F., Breton-Gorins, J., Trugman, G., Castro-Malaspina, H., Adrien, J.M., Bereziat, G., Lebret, M., and Caen, J.P.**, Studies on a new variant of the Hermansky-Pudlak syndrome: Qualitative, ultrastructural and functional abnormalities of platelet dense bodies associated with a phospholipase A defect, *Am. J. Hematol.*, 4, 387-399, 1978.

130. **Rendu, F., Lebebret, M., Nurden, A.T., and Caen, J.P.**, Detection of an acquired platelet storage pool disease in three patients with a myeloproliferative disorder, *Thromb. Haemost.*, 42, 794-796, 1979.

131. **Rendu, F., Maclouf, J., Launay, J.-M., Boinot, C., Levy-Toledano, S., Tanzer, J., and Caen, J.**, Hermansky-Pudlak platelets: Further studies on release reaction and protein phosphorylations, *Am. J. Hematol.*, 25, 165-174, 1987.

132. **Rendu, F., Nurden, A.T., Lebert, M., and Caen, J.P.**, Relationship between mepacrine-labelled dense body number, platelet capacity to accumulate 14C-5-HT and platelet density in the Bernard-Soulier and Hermansky-Pudlak syndromes, *Thromb. Haemost.*, 42, 694-704, 1979.

133. **Richards, J.G., and Da Prada, M.**, Uranaffin reaction: A new cytochemical technique for the localization of adenine nucleotide in organelles storing biogenic amines, *J. Histochem. Cytochem.*, 25, 1322-1326, 1977.

134. **Rudnick, G., Fishkes, H., Nelson, P.J., and Schudiner, S.**, Evidence for two distinct serotonin transport systems in platelets, *J. Biol. Chem.*, 255, 3638-3641, 1980.

135. **Russell, N.H., Keenan, J.P., and Frais, M.A.**, Thrombocytopathy associated with autoimmune haemolytic anemia, *Br. Med. J.*, 3, 604, 1978.

136. **Russell, N.H., Salmon, J., Keenan, J.P., and Bellingham, A.J.**, Platelet adenine nucleotides and arachidonic acid metabolism in the myeloproliferative disorders, *Thromb. Res.*, 22, 389-397, 1981.

137. **Sato, A., Tanabe, F., Ito, M., and Shigeta, S.**, Thiol proteinase inhibitors reverse the increased protein kinase C down-regulation and concavalin A cap formation in polymorphonuclear leukocytes from Chediak-Higashi syndrome (beige) mouse, *J. Leuko. Biol.*, 48, 377-381, 1990.

138. **Sjaastad, O.V., Blom, A.K., Stormorden, H., and Nes, N.**, Adenine nucleotides, serotonin, and aggregation properties of platelets of blue foxes (alopex lagopus) with the Chediak-Higashi syndrome, *Am. J. Med. Genet.*, 35, 373-378, 1990.

139. **Skaer, R.J., Flemans, R.J., and McQuilkan, S.**, Mepacrine stains the dense bodies of human platelets and not platelet lysosomes, *Br. J. Haematol.*, 49, 435-438, 1981.

140. **Starkey, J.R., Prieur, D.J., and Ristow, S.S.**, Natural killer cell activity in fawn-hooded rats, *Experientia*, 39, 308-310, 1983.

141. **Steinbrink, W.**, Uber eine neue granulationsanomalie der leukocytem, *Otsch Arciv. Klin. Med.*, 193, 577-581, 1948.

142. **Stuart, M.J.**, Platelet function in the neonate, *Am. J. Petiatr. Hematol. Oncol.*, 1, 227-234, 1979.

143. **Taylor, R.F., and Farrell, R.K.**, Light and electron microscopy of peripheral blood neutrophils in a killer whale affected with Chediak-Higashi syndrome, *Fed. Proc.*, 32, 822, 1973.

144. **Theuring, F., and Fiedler, J.**, Fatal bleeding following tooth extraction. Hermansky-Pudlak syndrome, *Otsch. Stomatol.*, 23, 52, 1973.

145. **Tobach, E., DeSantis, J.L., and Zucker, B.M.**, Platelet storage pool disease in hybrid rats. F1 fawn-hooded rats derived from crosses with their putative ancestors (Rattus norvegicus), *J. Hered.*, 75, 15-18, 1984.

146. **Tschopp, T.B., and Baumgartner, H.R.**, Defective platelet adhesion and aggregation on subendothelium exposed *in vivo* or *in vitro* to flowing blood of fawn-hooded rats with storage pool disease, *Thromb. Haemost.*, 38, 620-629, 1977.

147. **Tschopp, T.B., and Zucker, M.B.**, Hereditary defect in platelet function in rats, *Blood*, 40, 217-226, 1972.

148. **Weiss, H.J., and Ames, R.P.,** Ultrastructural findings in storage pool disease and aspirin-like defects of platelets, *Am. J. Pathol.*, 71, 447, 1973.

149. **Weiss, H.J., Chervenick, P.A., Zalusky, R., and Factor, A.,** A familial defect in platelet function associated with impaired release of adenosine diphosphate, *N. Eng. J. Med.*, 281, 1264-1270, 1969.

150. **Weiss, H.J., and Lages, B.,** Platelet malondialdehyde production and aggregation responses induced by arachidonate, prostaglandin-G2, collagen, and epinephrine in 12 patients with storage pool deficiency, *Blood*, 58, 27-33, 1981.

151. **Weiss, H.J., and Lages, B.,** The response of platelets to epinephrine in storage pool deficiency — evidence pertaining to the role of adenosine diphosphate in mediating primary and secondary aggregation, *Blood*, 72, 1717-1725, 1988.

152. **Weiss, H.J., and Rogers, J.,** Thrombocytopathia due to abnormalities in the platelet release reaction-studies on six unrelated patients, *Blood*, 39, 187-196, 1972.

153. **Weiss, H.J., Rosove, M.H., Lages, B.A., and Kaplan, K.L.,** Acquired storage pool deficiency with increased platelet-associated IgG. Report of five cases, *Am. J. Med.*, 69, 711-717, 1980.

154. **Weiss, H.J., Tschopp, T.B., and Baumgartner, H.R.,** Impaired interaction (adhesion-aggregation) of platelets with the subendothelium in storage pool disease and after aspirin ingestion, *N. Eng. J. Med.*, 293, 619-623, 1975.

155. **Weiss, H.J., Tschopp, T.B., Rogers, J., and Brand, H.,** Studies of platelet 5-hydroxytryptamine (serotonin) in storage pool disease and albinism, *J. Clin. Invest.*, 54, 421-432, 1974.

156. **Weiss, H.J., Turitto, V.T., and Baumgartner, H.R.,** Platelet adhesion and thrombus formation on subendothelium in platelets deficient in glycoproteins IIb-IIIa and storage granules, *Blood*, 67, 322-330, 1986.

157. **Weiss, H.J., Willis, A.L., Kuhn, D., and Brand, H.,** Prostagland E2 potentiation of platelet aggregation by LASS endoperoxides: Absent in storage pool disease, normal after aspirin ingestion, *Br. J. Haematol.*, 32, 257-272, 1976.

158. **Weiss, H.J., Witte, L.D., Kaplan, K.L., Lages, B.A., Chernoff, A., Nossel, H.L., Goodman, D.S., and Baumgartner, H.R.,** Heterogeneity in storage pool deficiency: Studies on granule-bound substances in 18 patients including variants deficient in alpha-granules, platelet factor 4, beta-thromboglobulin; and platelet-derived growth factor, *Blood*, 54, 1296-1319, 1979.

159. **Wey, H., Gallon, L., and Subbiah, M.T.,** Postaglandin synthesis in aorta and platelets of fawn-hooded rats with platelet storage pool disease and its response to cholesterol feeding, *Thromb. Haemost.*, 48, 94-97, 1982.

160. **White, J.G.,** Platelet microtubules and giant granules in the Chediak-Higashi syndrome, *Am. J. Med. Tech.*, 44, 273-278, 1978.

161. **White, J.G.,** Inherited abnormalities of the platelet membrane and secretory granules, *Hum. Pathol.*, 18, 123-139, 1987.

162. **White, J.G., Edson, J.R., Desnick, S.J., and Witkop, C.J.,** Studies of platelets in a variant of the Hermansky-Pudlak syndrome, *Am. J. Pathol.*, 63, 319-329, 1971.

163. **Wijermans, P.W., and Van Dorp, B.,** Hermansky-Pudlak syndrome: Correction of bleeding time by 1-desamino-8-D-arginine vasopressin, *Am. J. Hematol.*, 30, 154-157, 1989.

164. **Willis, A.L., and Weiss, H.J.,** A congenital defect in platelet prostaglandin production associated with impaired hemostasis in storage pool disease, *Prostaglandins*, 4, 147-153, 1973.

165. **Witkop, C.J., Hill, C.W., Desnick, S., Thies, J.K., Thorn, H.J., Jennkins, M., and White, J.G.,** Opthalmologic, biochemical, platelet, and ultrastructural defects in the various types of oculocutaneous albinism, *J. Invest. Dermatol.*, 60, 443-456, 1973.

166. **Witkop, C.J., Krumweide, M., and White, J.G.,** Reliability of absent platelet dense bodies as a diagnostic criterion for Hermansky-Pudlak syndrome, *Am. J. Hematol.*, 26, 305-311, 1987.

167. **Witkop, C.J., Quevedo, W.C., and Fitzpatrick, T.B.,** Albinism, in: *The Metabolic Basis of Inherited Disease*, 14th ed, Stanbury, J.B., Wyngaarden, J.B., Fredrickson, D.S., Eds., McGraw-Hill Book Company, New York, 1978, pp. 283-316.

168. **Witkop, C.J., Wolfe, L.S., Cal, S.X., White, J.G., Townsend, D., and Keenan, K.M.,** Elevated urinary dolichol secretion in the Hermansky-Pudlak syndrome: An indicator of lysosomal dysfunction, *Am. J. Med.,* 82, 463-470, 1987.

169. **Yamamoto, K., Sekiguchi, E., and Takatani, O.,** Abnormalities of epinephrine-induced platelet aggregation and adenine nucleotides in myeloproliferative disorders, *Thromb. Haemost.,* 52, 292-296, 1984.

170. **Zahavi, J., and Marder, V.J.,** Acquired "storage pool disease" of platelets associated with circulating antiplatelet antibodies, *Am. J. Med.,* 56, 883-890, 1974.

171. **Zucker-Franklin, D., Benson, K.A., and Meyers, K.M.,** Absence of a surface-connected canalicular system in bovine platelets, *Blood,* 65, 241-244, 1985.

INDEX

A

Abscess, 160
Acetylcholinesterase, 36
Acetylsalicylic acid, see Aspirin
β-N-Acetylglucosaminidase, 155, 162, 180
Acid hydrolase
  activation, 97
  congenital storage pool deficiency
    Chediak-Higashi syndrome, 158, 170, 171
    Hermansky-Pudlak syndrome, 153, 156, 180
Acid phosphatase, 155, 162, 180
Actin, 54, 56–57, 112, 113, see also Adenosine
    diphosphate
Adenosine diphosphate (ADP)
  alcoholism, 181, 182
  compartmentation in platelets, 1, 4, 97, 53–54
  epinephrine and, 126, 128–130
  human and porcine dense granules, 54–56
  5-hydroxytryptamine aggregation, 12
  idiopathic thrombocytopenia purpura, 180–181
  inositol trisphosphate induction, 81–82
  NMR studies, 52–61
  storage pool deficiency disorders, 149–151
    Chediak-Higashi syndrome, 158, 171, 173
    Hermansky-Pudlak syndrome, 19, 20, 153,
    155
Adenosine triphosphate (ATP)
  alcoholism, 181, 182
  compartmentation in platelets, 1, 9, 53–54
  human and porcine dense granules, 54–56
  NMR studies, 52–61
  polyphosphoinositide cycle, 98, 100, 104–107,
    109
  storage pool deficiency disorders, 149, 152, 155,
    158, 171, 173
Adenylate cyclase, 88, 130–131, 136
Adhesion, 153, 154
ADP, see Adenosine diphosphate
Adrenaline, 97
α-Adrenergic receptors
  epinephrine and, 128, 130–132, 135
  platelet function, 19, 117, 119–121
Afibrinogenemia, 124
Aggregation, see also Adenosine diphosphate;
    Adenosine triphosphate; Chediak-Higashi
    syndrome; Epinephrine; Hermansky-Pudlak
    syndrome; Storage pool disease
  epinephrine role, 117–119
  irreversible, 19, 121, 122–123
  nucleotides and, 67–69
  storage pool deficiency, 150–151, 153, 155

bovine, 163, 166–168
feline, 173, 175
immune-mediated, 180
mink and fox, 173
murine, 176
porcine, 178
Agnogenic myeloid metaplasia, 178, 180
Albinism, oculocutaneous, 19, 20, 154, 156, 157,
    see also Hermansky-Pudlak syndrome
Alcohol, abuse, 181
Alkaline phosphatase, 78
Alkenylacetyl glycerophosphocholine, 123–124
Amiloride, 136
Amine, see 5-Hydroxytryptamine
Angina pectoris, 138, 139
Animal models, see also Individual entries
  bovine Chediak-Higashi syndrome, 157, 158,
    160–171
  Hermansky-Pudlak syndrome, 156–157
  murine and feline Chediak-Higashi syndrome,
    157, 158, 171–177
Antibiotics, 125
Antibodies, 180
Antimycin A, 54, 58–61
Arachidonic acid
  epinephrine and, 124, 136–137
  polyphosphoinositide cycle and, 97, 100
  storage pool deficiency, 151–154
    Chediak-Higashi syndrome, 173
    Hermansky-Pudlak syndrome, 19, 20
Ascorbic acid, 160
Aspirin
  epinephrine and, 123–124, 139
  inositol phosphates, effect, 76
  irreversible aggregation by, 121–123
  storage pool deficiency, 19–21, 154
Atenolol, 139
ATP, see Adenosine triphosphate
Azaprostanoic acids, 125
Azide, 54, 58

B

Band broadening, 62–63
Baumgartner technique, 123
Beige mice, see Murine, Chediak-Higashi syn-
    drome model
Bernard-Soulier syndrome, 124
Bleeding, 21–22, 149–150, 154, 157, 178, see
    also Storage pool disease, acquired
Bleeding times
  membrane modulation, 138